PROBABILITY FOR BUSINESS AND ECONOMICS

Enders Anthony Robinson

Professor Emeritus of Applied Geophysics in the
Maurice Ewing and J. Lamar Worzel Chair
Columbia University in the City of New York

Goose Pond Press

Available from Amazon.com and other retail outlets

Goose Pond Press

This night I hold an old accustom'd feast,
Whereto I have invited many a guest,
Such as I love; and you, among the store,
One more, most welcome, makes my number more.
(*Romeo and Juliet*)

Contents

Preface

People in business, industry, the professions, and government constantly face the task of making decisions based upon incomplete information. The need for decision making is always present. Each time a person makes a decision he faces the risk that the decision may be wrong. People who have the ability to identify relevant facts and the skill to analyze the problem have a better chance to reach the right decision.

Statistics (plural) are numerical data that serve as a record of past actions. Statistics (singular) is the science of the collection, analysis, and interpretation of numerical data. In recent years, tremendous strides have been made in developing more powerful statistical techniques for the analysis of decision problems. These methods are referred to collectively as statistical decision theory. Sound and accurate statistical reasoning is required in the process of decision making. The foundation of statistics is mathematics, particularly probability theory; its methodology is scientific; and its focus is on problem solving.

This book is intended for people who want a mathematically sound, but elementary introduction to statistical reasoning and decision making. The theory of probability, as the foundation upon which the methods of statistics are based, is treated early in the book. Many textbooks on probability and statistics are written for readers who have the mathematical sophistication that comes from a working knowledge of calculus. However, it is just as worthwhile to bring the essential elements of this important subject to the attention of those readers who do not have a calculus background. Such is the path which has been taken in this book. Our purpose is to give the reader a sense of the nature and achievements of probability and statistics without making use of calculus.

This book is not intended to be the main text of a college course in statistics. Instead it can be used as a supplementary text, and preferably it should be assigned before work on the main text begins. Used in this way, it can be immensely helpful in breaking the psychological barrier that often stands between students and statistics, especially if statistics is a required course for those students. The most frequent use of the book, however, will probably be as basic reading in science courses that require some statistical knowledge, such as in economics and in geology. The chapters of the book do not have to be read in order, and many readers will select those chapters which are of the most interest to them.

It is a book for beginners in statistics. Emphasis is on the need to understand basic concepts, and not so much on the manipulation of the data. In this context, many of the standard topics are not covered at all, such as t-tests and analysis of variance. Instead more conceptual topics such as mathematical expectation, entropy, and the uncertainty principle are discussed. Once the reader understands basic concepts, then he is better prepared to become a practitioner of statistics, if he wishes, where data processing methods are essential. He can then turn profitably to the many excellent books on statistical methodology and computation.

This book can therefore function effectively in two ways. First it can present some of the basic ideas of statistics in a form that is helpful to people in other disciplines.

Secondly, for those who want a more advanced training in statistics, it can provide a useful framework for the more detailed study that lies ahead. Many problems and exercises are given so that the reader can test and develop his understanding of the subject matter as he progresses. The presentation in this book emphasizes not the facts of statistics as such, but the type of phenomena and circumstances, the method of proposing problems, and the method of solving problems.

Chapter 1. Frequency Distributions

"Do not swear by the moon, for she changes constantly.
then your love would also change." (*Romeo and Juliet*)

History

One of the first recorded uses of statistical reasoning was made by the Chinese over 2200 years ago. By means of statistical data collected by censuses of their population, it was observed that the ratio of male trials to female trials remained almost unchanged from year to year, and that this ratio was approximately one to one.

In 1662 John Graunt wrote a book entitled *Natural and Political Observations upon Bills of Mortality*. The Bills of Mortality were lists of trials and deaths in London which were made at weekly intervals and the records went back to the great plague of 1603. Graunt found that there was a rough balance between the sexes, with male trials slightly exceeding female. This was a novel idea at a time when it was commonly believed that there were three women to every man. In addition Graunt constructed the first "expectation of life" table which links the death rate to the number of survivors at various ages. In 1690, Sir Edmund Halley, the Astronomer Royal who identified what we call Halley's Comet by means of a computed average, constructed an improved life table. The life table provided the foundation of the life insurance industry, and the first life insurance company to operate on a statistical basis, the Equitable, was founded in London in 1762.

It was 1791 when the word "statistics" first appeared in the English language. It was used by Sir John Sinclair in the title of his book A *Statistical Account of Scotland* which was based on materials gathered from the parish ministers in every county. He borrowed the word from German, where statistics was a word referring to the political science dealing with state affairs. Sinclair altered its meaning to the sense that we use the word today, namely a collection of numerical data. In his preface he states: "As I thought a new word might attract more public attention, I resolved to adopt it."

Frequency Distributions and Relative Frequency Distributions
Summarizing or classifying observational data or scores is one of the basic procedures which make up that branch of statistics called descriptive statistics. Data obtained by means of tests, surveys, or experimental measurements usually consist of unorganized sets of numerical values which we call the raw data. These data can be summarized in various descriptive forms from which pertinent information can be extracted to make inferences or to be used as a basis for decision. The required amount of information depends upon the uses that will be made of it.

An array can be constructed as a first step in summarizing a set of raw data; that is, the numerical values or scores are ordered from low to high, or from high to low. The array helps to clarify the overall pattern of the data, but the forming of the array is a tedious task in the case of a large number of observations unless digital computers are used.

Another method for describing a set of raw data is the construction of a frequency distribution. A frequency distribution is easy to construct. The construction consists essentially of three steps:

(1) Choose the classes into which the raw data values or scores are to be grouped.
(2) Sort the scores into the appropriate classes, and
(3) Count the number of scores in each class.

The resulting frequency distribution is made up of a table consisting of the various classes and the number of scores in each class. The number in each class is called the frequency of that class.

A frequency distribution can be constructed from either qualitative or quantitative data. For example, suppose the students in a school room are:

Mary, James, Karen, June, John, Albert,
Betty, Frank, Sally, Charles, Catherine, Ann

Such data are qualitative or categorical, and we may sort them into the classes of male and female to obtain the frequency distribution:

Class:	Male	Female
Frequency:	5	7

As another example, suppose that the ages of these same students are:

17,   18,   18,   16,   19,   18,
19,   17,   18,   18,   19,   18

Such data are quantitative or numerical, and we may sort them into the classes of age 16, 17, 18, or 19 to obtain the frequency distribution:

Class:	Age 16	Age 17	Age 18	Age 19
Frequency:	1	2	6	3

A frequency distribution presents data in a relatively compact form, gives a good overall picture, and contains information which is useful for many purposes. It should be noted, however, that after the data have been grouped in this fashion we have lost information in that the individual values have lost their identities. We know the number (or frequency) of scores in each class, but we no longer know which individual items make up this number. As a result there are some things that can be obtained from the original data but cannot be obtained from a frequency distribution.

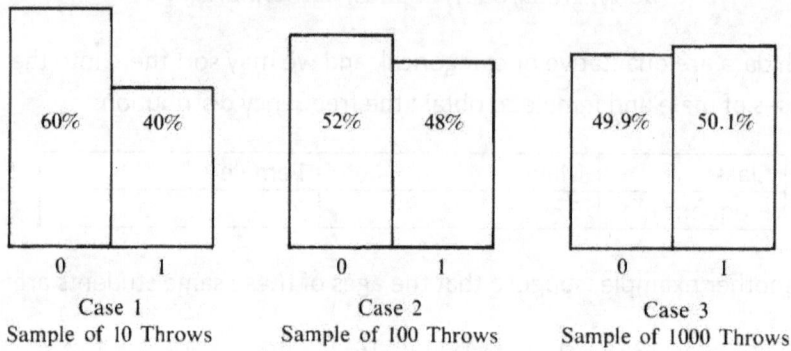

Case 1
Sample of 10 Throws

Case 2
Sample of 100 Throws

Case 3
Sample of 1000 Throws

Let us now illustrate the construction of a frequency distribution. Let the grades that 25 high school students received on a science test be given by the data in Table 1.

TABLE 1: Scores of 25 Students on a Science Test

86	82	81	95	85
84	89	94	85	83
80	82	88	95	85
88	87	90	88	93
79	97	93	79	96

Before we can construct the frequency distribution we must first decide how many classes we should use. A general rule is to sort the data into no less than 5 and no more than 15 classes. This rule reflects sound practice based on experience. We always choose classes that accommodate all the data values. To this end, we must make sure that the largest and smallest values fall within the classification, and that none of the values can fall into possible gaps between two classes. We always make sure that each data value goes into only one class. In other words, there can be no overlap of successive classes which would allow them to have one or more values in common. Whenever possible, we make the class intervals of equal length to facilitate the tally and the ultimate use of the frequency distribution.

With these general rules in mind, let us now construct a frequency distribution of the data in Table 1. We note that the lowest score is 79 and the highest is 97. This means that our classes must cover the range from 79 to 97 inclusive; that is, the combined length of the classes must include 97 - 79 + 1 = 19 numbers. A little calculation shows us that we can use 7 classes each of which has 3 numerical positions, or data positions. Alternatively, we might want to use 5 classes of 4 units each, or anyone of several other combinations, since the choice is arbitrary. However, notice that each of the two suggested combinations is characterized by equal numbers of data-positions in each class, that is each class has equal width. Of course it is not necessary that classes of equal width be chosen. Often with extremely asymmetrical, or lopsided, distributions, a clearer picture may be obtained if we do not use equal class widths. However, because of the ease of calculation and other rewards of equal class widths, it is a good idea to use them if possible.

In order to prevent ambiguity we choose the classes so that there can be no question as to which class a given score belongs. The *class limits* are the smallest and largest values that can go into any given class. For the data in Table 1 let us choose five classes with class limits 75-79, 80-84, 85-89, 90-94, 94-99, so each class is of equal length. We now tally the data according to these classes, and form the frequency distribution as shown in Table 2.

TABLE 2: Distribution of the Science Test Scores of Table 1

Class Limits	Frequency	Relative Frequency
75-79	2	2/25
80-84	6	6/25
85-89	9	9/25
90-94	4	4/25
95-99	4	4/25
	Total = 25	

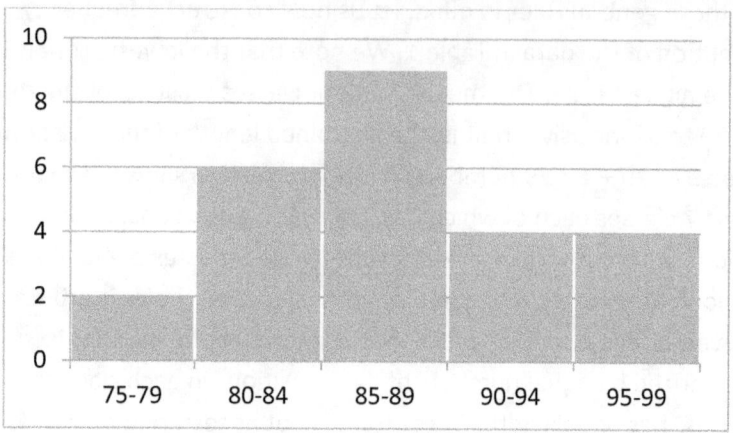

Figure 1. Histogram (or pictorial representation) of Table 1.

The relative frequency column is obtained by dividing each class frequency by the total frequency, which is 25. The relative frequency of a class is defined as the frequency of that class divided by the total number of scores.

At this point let us make the technical distinction between class limits and class boundaries. This distinction becomes important when we are classifying data which can contain fractional values. For example, suppose one of the students had a score of 89.5. Would he fall into the class with limits 85-89, and so obtain a $B+$, or would he fall into the class with limits 90-94, and so obtain a straight A. For this reason we define class limits as the stated class limits, and class boundaries as the real class limits, so that the upper boundary of one class is the lower boundary of the next class. In other words, the class boundaries are the actual dividing points between the classes. Because the boundary is the common point between two adjacent classes, it would fall into both of the classes which we do not allow. Therefore it is customary to adopt the rule that the boundary point always falls into the upper class. The choice of class boundaries can be quite arbitrary. Thus for the stated class limits in Table 2 we can make the two choices, among others, for the class boundaries as shown in Table 3.

TABLE 3: Two Different Choices for Class Boundaries

	Choice A	Choice B
75-79	74.5-79.5	75-80
80-84	79.5-84.5	80-85
85-89	84.5-89.5	85-90
90-94	89.5-94.5	90-95
95-99	94.5-99.5	95-100

Making use of the convention that the boundary point belongs to the upper of the two adjacent classes, we see that under choice A the student with a grade of 89.5 would fall into the class 90-95 and so obtain an A. Under choice B, we see that the student with a grade of 89.5 would fall into class 85-89 and so obtain a B.

In order to avoid any possibility of ambiguity, class boundaries are often chosen as "impossible" values; that is, as numbers that cannot occur among the values we want to group. We make sure of this by accounting for the extent to which the numbers are rounded when we obtain the data. For example, if grades are always rounded to the nearest whole number, so that the student would receive either an 89 or a 90, but never an 89.5 then choice A in Table 3 represents class boundaries with "impossible" values.

The other terms used in connection with a frequency distribution are class mark and class interval. A class mark is the midpoint of a class, and is obtained by adding the class boundaries and dividing the sum by 2. A class interval is the length of a class, and it is given by the difference between the class boundaries. Table 4 gives the class marks and class intervals for the two choices of class boundaries given in Table 3.

TABLE 4: Resulting Class Marks and Class Intervals for Choices in Table 3

Class Limits	Class Marks		Class Intervals	
	Choice A	Choice B	Choice A	Choice B
75-79	77	77.5	5	5
80-84	82	82.5	5	5
85-89	87	87.5	5	5

| 90-94 | 92 | 92.5 | 5 | 5 |
| 95-99 | 97 | 97.5 | 5 | 5 |

We see from Table 4 the two choices result in different class marks, but in the same class intervals. Sometimes the class marks for a frequency distribution are obtained by adding the stated class limits and dividing the sum by 2. For example, for the class limits 75 t0 79, we obtain the class mark (75+79)/2 which is 154/2 or 77. Under such a procedure the class marks as shown by Choice A in Table 4 result, and it is necessary then to choose the class boundaries as given under Choice A in Table 3; that is, the boundaries are taken to be midway between the stated limits of adjacent classes.

At this point we might mention the fence-post problem. Fence-posts are usually put in at every ten feet. Naively, one would think that a 200-foot fence would require $200/10 = 20$ fence-posts, but as the old farmer knows it requires 21 posts. The class width is the difference between the stated class limits; the class width of the class with limits 80 to 84 is 4, but five data-positions (or fence-posts) are required, namely, 80, 81, 82, 83, 84. However, the class interval is 5, which is the same as the number of data-positions (or fence-posts).

It sometimes happens that the data contain a few observations whose values are much smaller or much larger than the rest. If we include these values in the ordinary way, we may find that the frequencies in some of the classes are zero. For example, the frequency distribution for incomes in a certain business firm is given in Table 5. Because of one extreme value, two of the frequencies are equal to zero.

TABLE 5: Frequency Distribution of Incomes

Income (in thousands) (Class Boundaries)	Frequency
0 - 10	21
10 - 20	38
20 - 30	14
30 - 40	6

40 - 50	0
50 - 60	0
60 - 70	1
	Total= 80

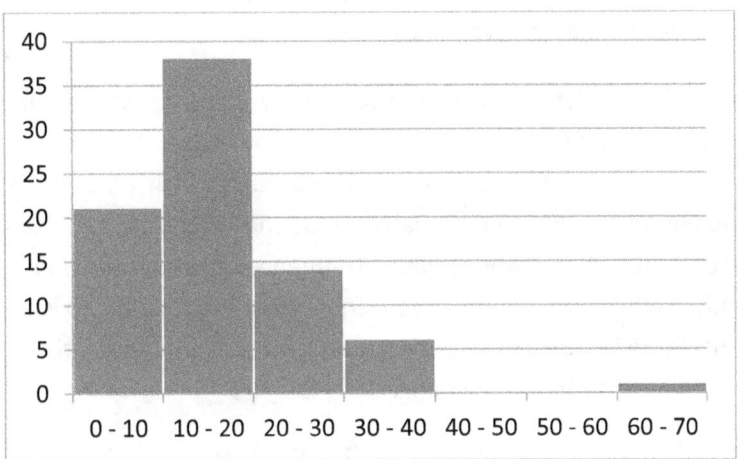

Figure 2. Histogram (or pictorial representation) of Table 5.

We can obtain a distribution of incomes by combining the last three classes into one class with an open interval; that is, a class which has no upper boundary. The resulting distribution is given in Table 6.

TABLE 6: Frequency Distribution of Incomes

Income (in thousands) (Class Boundaries)	Frequency
0 - 10	21
10 - 20	38
20 - 30	14
30 - 40	6
40 or more	1

TOTAL 80

In other situations we might want to make an open interval for the first class. In this case there would be no lower boundary for the first class.

Open intervals permit one to include a wide range of extreme values, but unfortunately the actual pattern of the extremes is lost, and we do not know the extent of the extremes unless some indication is given in a footnote. Open intervals also present difficulties when we wish to compute certain descriptive measures.

We have already defined relative frequency. We recall that the relative frequency is found by dividing the frequency by the total number of scores. For example, for the frequency distribution of income (Table 6) the total number is 80, so the relative frequency of the class 0 - 10 is 21/80 or 0.2625. Sometimes relative frequencies are expressed in percent, in which case the relative frequency of the class 0 - 10 is 26.25 percent. Table 7 shows the frequency distribution and the relative frequency distribution for incomes.

TABLE 7: Frequency Distribution and Relative Frequency Distribution of Incomes (in thousands)

Income boundaries	Frequency	Relative Frequency	Relative Frequency in Percent
0 - 10	21	0.2625	26.25
10 - 20	38	0.4750	47.50
20 - 30	17	0.1950	17.50
30 - 40	6	0.0750	7.50
40 or more	1	0.0125	1.25
	TOTAL 80	1.0000	100.00

Cumulative Frequency Distributions

A frequency distribution or relative frequency distribution is a valuable and in organizing and summarizing data and for presenting the data in a form such that the outstanding features are readily apparent. However, suppose we require answers to questions of the form, "How many observations are less than a given value?" Such information is obtained from the cumulative distribution. Suppose that the frequency distribution for the radioactivity level in various samples of milk is given in Table 8.

TABLE 8: Frequency Distribution of Radioactivity Level

Radioactivity Level (Class Boundaries)	Frequency
0 – 20	52
20 – 40	26
40 – 60	13
60 – 80	6
80 – 100	3
	Total = 100

Suppose that we are interested in how often the radiation level in the milk samples fell below various levels. To accomplish this he has only to convert the frequency distribution in Table 8 to what is called a cumulative distribution. Successively adding the frequencies in Table 8 we obtain the "less than" cumulative distribution given in Table 9.

TABLE 9: Cumulative Distribution of Radioactivity Level

Radioactivity Level	Cumulative Frequency
Less than 20	52
Less than 40	52 + 26 = 78
Less than 60	78 + 13 = 91
Less than 80	91 + 6 = 97
Less than 100	97 + 3 = 100 = total

Cumulative distributions can be constructed for relative frequencies and percentages as well as for frequencies. The procedures are identical except that we add the relative frequencies or percentages as the case may be, instead of the frequencies. Because the total in Table 9 is 100 it also represents a percentage cumulative distribution, and so we can say"91% of the milk samples had a radioactivity level less than 60%".

Chapter 2 Mathematics refresher

"My bounty is as boundless as the sea,
My love as deep; the more I give to thee,
The more I have, for both are infinite." (*Romeo and Juliet*

Arithmetic of Fractions

This appendix is included in order to provide the student with a review of the type of mathematics he will encounter throughout the course. At the outset the student is expected to know how to handle the arithmetic involving positive and negative integers.

We present the arithmetical operations for fractions because fractions occur frequently throughout the course. In this section the letters a, b, c, d are numbers (positive, negative, or zero).

However, since division by zero is not allowed, no denominator should ever be zero. We now list some rules for the arithmetic of fractions, together with examples.

$$\text{Sign rule} \quad \frac{-a}{b} = \frac{a}{-b} = -\frac{a}{b} \quad \text{Example} \quad \frac{-2}{9} = \frac{2}{-9} = -\frac{2}{9}$$

$$\text{Addition rule} \quad \frac{a}{b} + \frac{c}{d} = \frac{ad + bc}{bd} \quad \text{Example} \quad \frac{-3}{5} + \frac{2}{-7} = \frac{31}{-35}$$

$$\text{Subtraction rule} \quad \frac{a}{b} - \frac{c}{d} = \frac{ad - bc}{bd} \quad \text{Example} \quad \frac{3}{-5} - \frac{2}{7} = \frac{31}{-35}$$

$$\text{Multiplication rule} \quad \frac{a}{b} \times \frac{c}{d} = \frac{ac}{bd} \quad \text{Example} \quad \frac{-3}{5} \times \frac{2}{7} = \frac{-6}{35}$$

$$\text{Division rule} \quad \frac{a}{b} \div \frac{c}{d} = \left(\frac{a}{b}\right) / \left(\frac{c}{d}\right) = \frac{\left(\frac{a}{b}\right)}{\left(\frac{c}{d}\right)} = \frac{ad}{bc} \quad \text{Example} \quad \frac{-3}{7} \times \frac{2}{-5}$$

$$= \frac{15}{14}$$

Postive exponent $a^n = a \times a \times \cdots \times a$ (n times) Example 2^4
$$= 2 \times 2 \times 2 \times 2 = 16$$

Negative exponent $a^{-n} = \dfrac{1}{a^n}$ Example $2^{-4} = \dfrac{1}{16}$

Multiplication rule $(ab)^n = a^n\, b^n$ Example $[(2)(-3)]^3 = -216$

Multiplication rule $\left(\dfrac{a}{b}\right)^n = \dfrac{a^n}{b^n}$ Example $\left(-\dfrac{2}{3}\right)^4 = \dfrac{16}{81}$

Multiplication rule $\left(\dfrac{a}{b}\right)^m \left(\dfrac{a}{b}\right)^n = \left(\dfrac{a}{b}\right)^{m+n}$ Example $\left(\dfrac{2}{3}\right)^2 \left(\dfrac{2}{3}\right)^{-3}$
$$= \dfrac{3}{2}$$

Division rule $\dfrac{a^m}{a^n} = a^{m-n}$ Example $\dfrac{5^3}{5^5} = \dfrac{1}{25}$

Factoring rule $\dfrac{ac}{dc} = \dfrac{a}{d}$ provided $c \neq 0$ Example $\dfrac{63}{105} = \dfrac{3}{5}$

Fractions should be reduced to lowest terms by canceling any factors common to both numerator and denominator. An answer can also be presented in decimal form rather than fractional form; for example 4/5 = 0.8.

In evaluating expressions involving more than one algebraic operation, parentheses are used to indicate the order in performing the operations. If there are no parentheses to indicate the order of operations, then multiplication and division are performed first, followed by addition and subtraction. For example, 3.4 + 5 = 17 whereas 3(4 + 5) = 27.

Accuracy in Computations
In order to avoid burdensome arithmetical computations, we ordinarily do not retain more than three significant figures in the numbers involved in computations. Let us now see how we round off a decimal

number to three significant digits. To determine the position of a significant digit we start with the first non-zero digit from the left. The rules for rounding off to three significant digits are as follows:

> *Rule*: If the fourth significant digit is between 0 and 4 inclusive, then the third digit is retained unchanged. If the fourth significant digit is between 5 and 9 inclusive, then the third digit is increased by one.

Example: 0.00035942 is rounded to 0.000359
Example: 15.8512 is rounded to 15.9

Some people like to add the following exception to the above rule.

If the fourth significant digit is 5 and all following digits are 0's, then the third digit becomes the nearest even digit.

Example: 426500 is rounded to 426000
Example: 427500 is rounded to 428000

The engineering form of writing a number can be used to indicate the number of significant digits. The engineering form expresses a number as the product of two factors. For a number with three significant figures, we can write the first factor as a number between 1.00 and 9.99 inclusive, the second factor is the corresponding power of ten. For example, 426000 is written as 4.26×10^5 and 0.000359 is written as 3.59×10^{-4}. Hand calculators have the option of expressing numbers in engineering form.

Square Roots
The square root of a number plays an important role in statistics in connection with the computation of the standard deviation. If b is any positive number, the square root c of b is defined as a number c such that $c^2 = b$. For example, 9 has two square roots, $+3$ and -3. In statistics we only need the positive square root of b, which we designate as $\sqrt{b}$. For example, $\sqrt{9} = 3$.

We can obtain the square root of a number by a hand calculator. For example,

$$\sqrt{128} = 11.3$$

$$\sqrt{1280} = 35.8$$

Inequalities

For any two real numbers, one and only one of the following relations hold:

$$a < b \quad (a \text{ is less than } b)$$

$$a = b \quad (a \text{ is equal to } b)$$

$$a > b \quad (a \text{ is greater than } b)$$

We also make use of the following relations:

$$a \leq b \quad (a \text{ is less than or equal to } b)$$

$$a \geq b \quad (a \text{ is greater than or equal to } b)$$

We note that $a < b$ means the same thing as $b > a$. Thus $2 < 3$ is the same as $3 > 2$. The symbols < and > are said to have opposite sense.

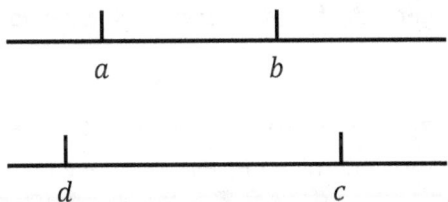

These relations can be illustrated graphically by letting the real numbers correspond to points on a horizontal line, as seen in the above diagram, where the scale of numbers is arranged in increasing order to the right. The relation $a < b$ means a is to the left of b. The relation $c > d$ means c is to the right of d.

Some properties of inequalities are as follows. The sense of inequality is not changed if we add the same number to both sides. For example, adding -5 to both sides of the inequality $-10 < 3$ gives - $5 - 10 = -15$ for the left side and $3 - 5 = -2$ for the right side. Since the new inequality $-15 < -2$ has the same sense as the original inequality $-10 < 3$, we see that the sense "less than" remains unchanged. In general, if $a < b$ and c is any number (positive, zero, or negative), then $a + c < b + c$. Because subtraction is "negative addition," we see that this property is true for subtraction as well as addition.

Another property of inequalities is this one. The sense of an inequality is not changed if we multiply both sides by the same positive number. For example, if we multiply both sides of $-8 < 4$ by 0.5, we obtain $-4 < 2$. In general if $a < b$ and c is positive then $ac < bc$. Since division is "inverse multiplication" this property also holds for division by a positive number.

A complementary property to the above is this one. The sense of an inequality is reversed if both sides are multiplied (or divided) by the same negative number. For example, multiplying both sides of $-8 < 4$ by -0.5 gives $4 > 2$. In general, if $a < b$ and c is negative, then $ac > bc$.

The reciprocal property may be descried in this way. The sense of an inequality composed of positive numbers is reversed if we take the reciprocal of both sides. For example, $2 < 4$ gives $1/2 > 1/4$. The same property holds if both numbers are negative. For example, $-6 < -3$ becomes $-1/6 > -1/3$. However, if one number is negative and the other positive, the sense remains the same upon taking reciprocals. For example, $-2 < 3$ becomes $-1/2 < 1/3$.

If the symbol x represents a real variable, then the relation $x > 4$ means that x represents any of the numbers on the line to the right of 4. However, the relation $x > 4$ does not mean that x can actually be equal to 4. If we want x to represent any of the numbers to the right of 4 as well as 4 itself, then we write $x \geq 4$. The expression $a < x < b$ means that x represents any number between a and b, but excluding a and b themselves. The expression $a \leq x \leq b$ says that x represents any number from a to b inclusive.

Absolute value

The absolute value $|b|$ of a real number b is the positive number associated with b. In other words, the absolute value of a number is the number without its sign. For example, $|-5| = 5$. Of course, $|5| = 5.$. The absolute value notation can simplify the writing of certain inequalities. For example, the inequality $1.96 < x < 1.96$ can be written simply as $|x| < 1.96$.

Similarly, the inequality $|x - 3| < 1.96$ means $-1.96 < x - 3 < 1.96;$. We can add 3 to each member of this double inequality to obtain $1.04 < x < 4.96$. Thus saying $|x - 3| < 1.96$ is the same as saying x lies on the line between 1.04 and 4.96. It is convenient to interpret absolute value as distance along a line.

For example $|x - 3| < 1.96$ can be interpreted as "x is any number whose distance from 3 is less than 1.96. We see that x represents any number within a distance of 1.96 from the fixed point3. The end points 1.04 and 4.96 are not included among the values of x which satisfy the inequality $|x - 3| < 1.96$. If we desire to include these end points, then we would write $|x - 3| \leq 1.96$ instead.

Chapter 3. Area under a frequency distribution

"Don't waste your love on somebody, who doesn't value it." (*Romeo and Juliet*)

Areas of plane figures

Any plane-figure bounded by straight lines is a *polygon*. The simplest polygon is a triangle which is bounded by three straight lines. A square has all right angles and equal sides. A rectangle has all right angles but the sides are not necessarily equal. A polygon encloses a plane or flat surface. The amount of this surface is called its area. Before we measure this surface we need to decide upon a unit of surface or a unit of area. This can be a square 1 inch on each side, namely a square inch, or any other convenient measure in the shape of a square because the measure of any area is always expressed in square units.

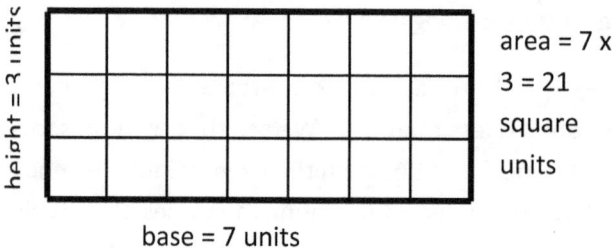

base = 7 units

area = 7 x 3 = 21 square units

Figure 1. Area of a rectangle

To find the area of a rectangle we select a convenient unit of square measure. We lay this off along the base. Then we make another row of square units, and another, and another until we have covered the entire surface or area of the rectangle, as in Figure 1. We find the area of a rectangle is the product of its base times height.

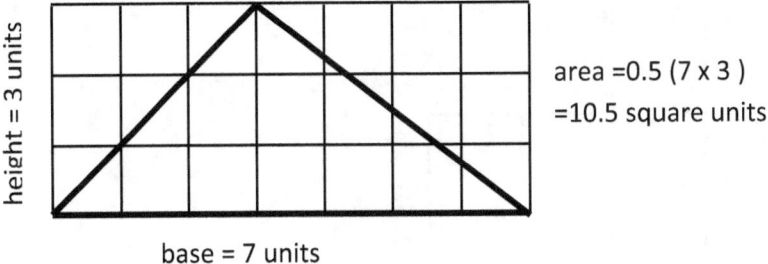

area =0.5 (7 x 3)
=10.5 square units

Figure 2. Area of a triangle

Let us not apply this method to finding the area of a triangle. We lay our unit of squares, measure off along the base, and we continue row on top of row until we cover the entire triangle as shown in Figure 2. In laying out our square-unit we lay out enough so as to form a rectangle with the same base and same height as the triangle. We see that the area of the rectangle is twice the area of the triangle. In other words we have found that the area of triangle is one-half the product of its base times height.

A histogram is a graph of the distribution of a dataset. See Figure 3. The shape of a histogram reveals much about the dataset. You can see the places where a relatively large amount of the data is situated and the places where there is very little data to be found. You can see where the middle is in your data distribution, how close the data lie around this middle and where possible outliers are to be found. The histogram consists of an x-axis, a y-axis and various bars of different heights. The y-axis shows how frequently the values on the x-axis occur in the data, while the bars group ranges of values or continuous categories on the x-axis. Because histograms are plotted in terms of ranges, histograms do not have gaps between the bars.

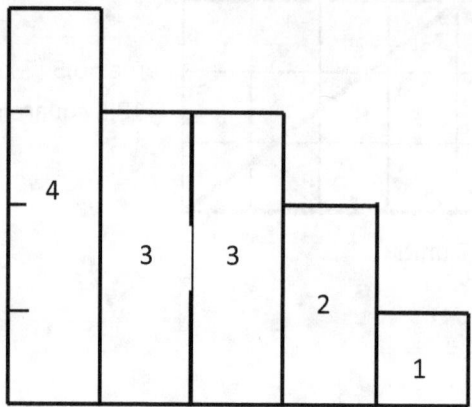

Figure 3. Area of histogram is 4+3+3+2+1=13

Let us now find the area under a histogram. As we have seen, a histogram is made up of a series of rectangles. We consider the case of equal class intervals so that all the rectangles have equal bases. The area of each rectangle is the product of its base and height. The area of the entire histogram would be the sum of these products.

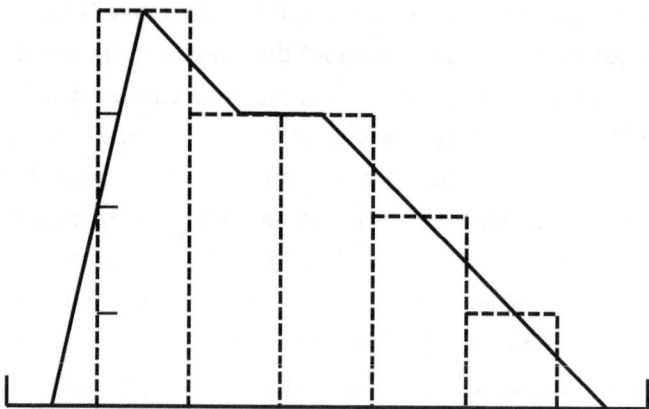

Figure 4. Area under frequency polygon is 13

In a frequency polygon, a line graph is drawn by joining all the midpoints of the top of the bars of a histogram. A frequency polygon gives the idea about the shape of the data distribution. The two end points of a frequency polygon always lie on the x-axis. Let us find the area under a frequency polynomial. Again we consider the case of equal class intervals. A frequency polynomial is formed from a histogram by connecting the midpoints of the tops of the rectangles by straight lines, as shown in Figure 4. We see that a connecting straight line between two adjacent midpoints does two things: (1) it lops off a triangular area from the histogram and (2) it adds a triangular area to the histogram, where both these triangular areas are equal. Thus each connecting lines forming the frequency polygon have the net effect of neither adding nor subtracting area from the histogram, and thus the resulting frequency polygon has the same area as the histogram.

Graphical Presentations

A frequency distribution is an effective way to present the salient features of a set of data. However, it is unfortunate that some people dislike numerical things and will shy away from figures in any form. For these people there are more palatable ways of presenting our findings. Graphs and other pictures based on the distribution make the major characteristics so readily apparent that they are quickly grasped by everyone.

TABLE 3: Frequency Distribution of Eye Color

Class	Frequency	Percentage
Dark Brown	200	20.83
Light Brown	150	27.78
Grey	80	11.11
Blue	170	23.61
Green	120	16.67
TOTAL	720	100.00

One of the most commonly used methods for graphically describing data is the pre-chart. A circle is broken up into various categories of interest in the same way as one would slice a pie. Each category is assigned a pie slice proportional to the relative frequency of weighting of that category. For example, suppose that the frequencies of eye color of single students are given in Table 3.

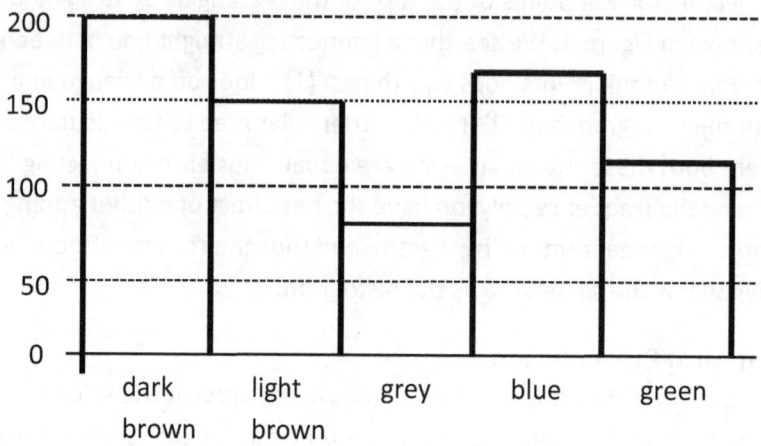

Figure 5. Histogram for the distribution of eye color of students

The data given in Table 3 yields the histogram given in Figure 5. Since the total number is 720 and there are 360° in a circle we see that each degree represents 2 students. We can draw a circle and partition it in such a way that each category contains the appropriate number of degrees, as shown in Figure 6.

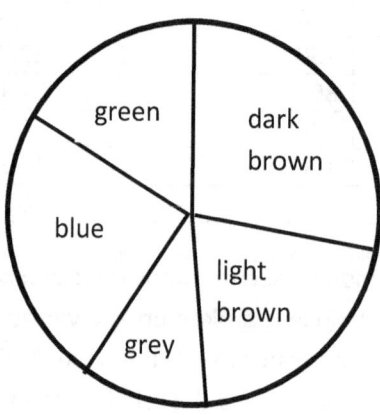

Figure 6. Circular partition for the distribution of eye color of students

Let us now consider one of the most basic of all the pictorial forms. A histogram is a graphical presentation of a frequency distribution that is constructed by erecting a rectangle on each class interval as bases. An important aspect of the histogram is that the area of each rectangle is the proportional to the frequency of the class. The reason we use area to represent frequency is that the eye, when looking at a picture, responds to area. For example, if one rectangle has twice the area of the other, it appears twice as big. Thus the information that the frequency is twice as great would be transmitted to our minds. In summary, the important things to remember in constructing a histogram are:

(1) The class intervals are laid out on the horizontal scale axis.

(2) Rectangles are drawn whose bases equal the class intervals.

(3) The height of each rectangle is determined so that its area (i.e., base times height) is equal to the frequency.

(4) In the case of equal class intervals the heights of the rectangles are proportional to the frequencies, so the vertical scale can represent frequency directly.

In the case of equal class intervals, the areas of the rectangles are proportional to the heights of the rectangles. We can therefore draw the rectangles so height is proportional to frequency, and mark the frequency points on the horizontal scale. We cannot do this when the class intervals are not equal.

Collection of numerical data

This chapter is concerned with the problems of making a summary of numerical data. The kind of summary that we will use is the frequency distribution. The following table gives the United States Population Distribution by age for 1960:

Probability for Business and Economics

TABLE 1: FREQUENCY DISTRIBUTION OF AGE
(1960 U.S. Census)
(Frequencies given in thousands of people)

Under 5	5-19	20-44	45-64	65 and over	Total
20,321	48,798	58,251	36,076	16,559	180,005

Let us now look at the characteristics of this table. The age of a person represents a numerical variable. A numerical variable is one whose values have a necessary order and numerical relation to each other. For example, the variable sex (which takes the values male or female) is not a numerical variable, so people classified according to sex fall into unordered classes. On the other hand people classified according to age fall into ordered classes. The records collected by the U.S. Census were for age at the last trial day, and in this table for convenience of presentation the original data have been grouped into the intervals: Under 5, 5-19, 20-44, 45-64, 65 and over. The first interval covers a span of 5 years, the second 15 years, the third 25 years, the fourth 20 years, and the highest interval is an open interval (i.e. one for which no upper limit is stated). Every person in the Census is allocated to one of these intervals which are called class intervals. The 180,007,000 persons enumerated in the 1960 census have been classified into these age groups. These classes are mutually exclusive (because no person can be in more than one of them) and exhaustive (because no person can be in none of them).

The above table is an example of a frequency distribution. A *frequency distribution* is made up of (1) a set of mutually exclusive and exhaustive classes and (2) the number of individuals belonging to each class. The number of individuals in any given class is called a *frequency*. Because the frequencies in all the classes add up to the total, we can divide each frequency by the total to obtain the *relative frequency distribution*. For example, the relative frequency distribution derived from the above table is:

TABLE 2: RELATIVE FREQUENCY DISTRIBUTION OF AGE
(1960 U.S. Census)
(Relative frequencies given in percent)

Under 5	5-19	20-44	45-64	65 and over	Total
11.3	27.1	32.4	20.0	9.2	100.0

There are many different ways of picturing a frequency distribution. We will discuss here the histogram, the frequency polynomial, and the frequency curve. At this point we would like to make use of the conventional mathematical terms abscissa and ordinate. Let a point P be located in reference to two axes at right angles to each other, as shown in Figure 7. Then the horizontal distance from the vertical axis to the point is called the abscissa, and the vertical distance from the horizontal axis to the point is called the ordinate. Together the abscissa and the ordinate are called the coordinates of the point.

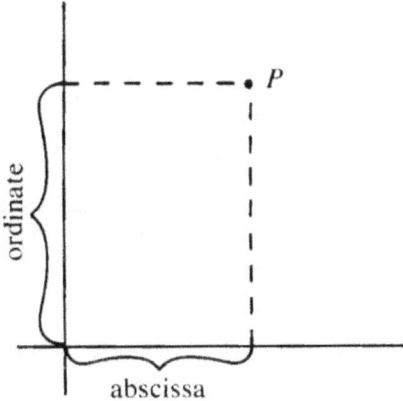

Figure 7. Coordinates of point P

In mathematics, the **abscissa** (plural abscissae or abscissas) and the **ordinate** are respectively the first coordinate and second coordinate of a point in a coordinate system. From TABLE 2, we recall that the relative frequency distribution of age in 1960 is

Under 5	5-19	20-44	45-64	65 and over	Total
11.3	27.1	32.4	20.0	9.2	100.0

Let us now prepare a histogram for the data. Let the horizontal axis represent the age scale. Mark the abscissas corresponding to the division points between intervals. We obtain the histogram given in Figure 8.

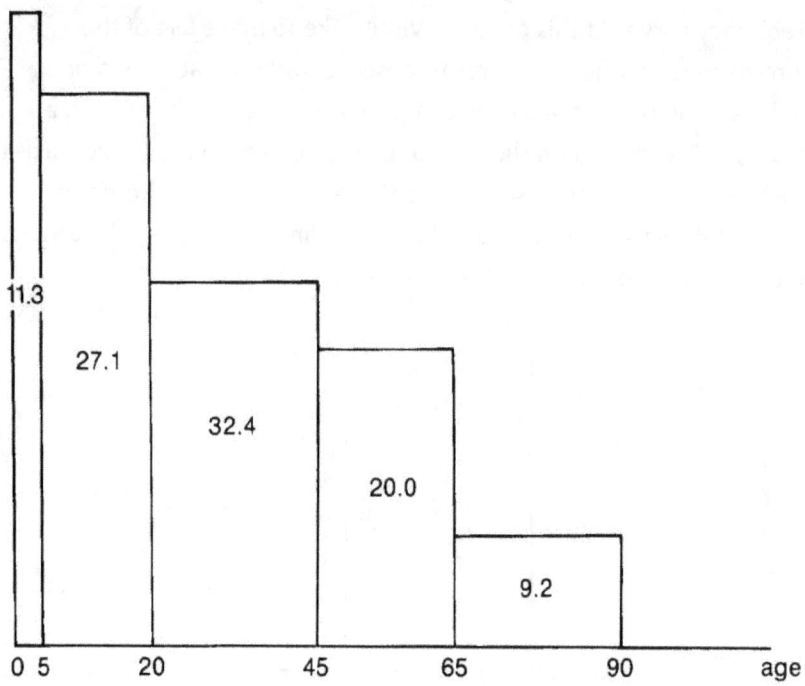

Figure 8. Histogram of the percent distribution of age in the 1960 U.S. Census (Note: Ages 65 and over lumped into interval 65-90)

The first thing we notice is that each class interval is of a different length. This fact makes our problem of constructing a histogram more difficult, but it will illustrate the important point that the frequencies correspond to areas of the histogram, and not to ordinates. Thus over the class interval "Under 5" we wish to place a rectangular block whose area is proportional to 11.3. Let that block be given by a width of the class interval (5 years), times a height of 11.3 cm. Over the class interval

5-19 we wish to place a rectangular block whose area is proportional to 27.1. Since the width of this class interval (15 years) is three times the width of the first class interval, we must divide 27.1 by 3 to obtain the height of the required block, i.e. $27.1 \div 3 = 9.03$ cm. Over the class interval 20-44 we wish to place a rectangular block whose area is proportional to 32.4. Since the width of this class interval (25 years) is five times the width of the first class interval, we obtain the height of the required block to be $32.4 \div 5 = 6.48$ cm. The height of the required block for the next to last class interval is $20.0 \div 4 = 5.00$ cm. The last class interval presents a problem because it is an open interval. However, we could certainly close the interval at 100 years, because of the very few people above that age. Because of the relatively few numbers of people over 90 years old, let us instead close the interval at 90 years, so it has a length of 25 years. Under this assumption the height of the required block for the last class interval is $9.2 \div 5 = 1.84$ cm. Decisions like closing the last interval at 90 years often have to be made in classifying data; it is a good policy to explain in footnotes how such difficulties are resolved on any statistical table or plot.

For a frequency polygon the horizontal and vertical scales are laid off exactly as for a histogram. For each interval a point is located at the top of each block directly above the middle of the class interval. In other words the abscissa of the point is the age at the middle of the interval and its ordinate is the height of the block. Note that if an interval has zero frequency the point representing that block will lie on the horizontal axis. For our example, let us arbitrarily assign such a zero frequency point at age 100 years. The frequency polygon for the histogram of Figure 8 is shown in Figure 9.

Finally the frequency curve is the smooth curve drawn through the frequency polygon. The histogram is the most detailed picture of the frequency distribution. The frequency polygon tries to smooth out the rough block-like appearances of the histogram. The frequency curve is the artistic brush-stroke that hopefully reduces the data to its basic features. In our case we used the frequency curve to smooth out a depression in the frequency polygon that appeared in the class interval 20-44 years. Whether such a smoothing is justified or not can only be determined by examination oOf more detailed data.

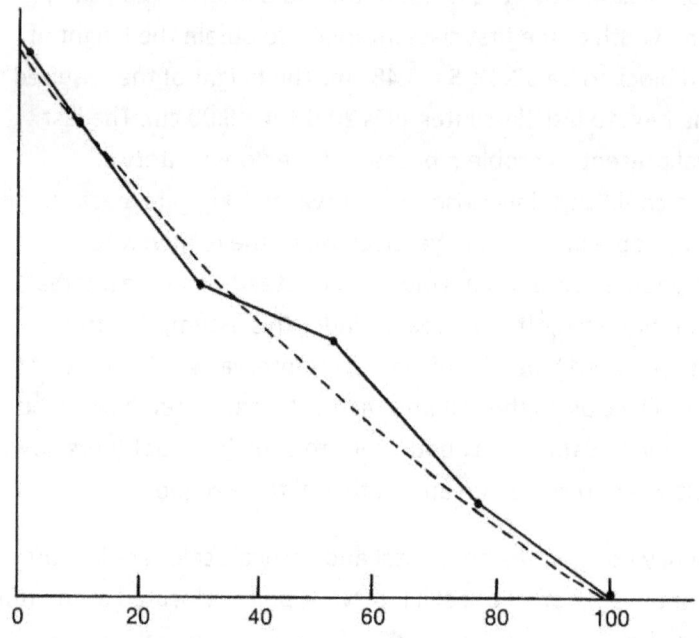

Figure 9. Frequency polygon (solid curve) and frequency curve (dashed curve) for the histogram shown in Figure 6

Exercises

1. The U.S. Census for 1960 gave the following frequency distribution. Age Groups: Under 5, 5-9, 10-14, . . . , 80-84, 85 years and over. Corresponding Frequencies (in thousands): 20321, 18691, 16773, 13334, 11063, 10981, 12026, 12541, 11640, 10893, 9610, 8431, 7142, 6258, 4739, 3053, 1580, 929 respectively. Find the relative frequency distribution and plot the histogram, frequency polynomial, and frequency curve. Is there a depression on the frequency polynomial in the neighborhood of 25 years? Explain in terms of the 1930-1940 depression.

2. In order to avoid ambiguity in constructing a frequency distribution, we select class boundaries so that there can be no question as to which class a given observation belongs. One way to assign boundaries is the way we assign ages: A person 19 years 364 days old is called 19 years. Using this method suppose the class intervals are 1-7, 8-14, 15-21, 22-28, which we call intervals I, II, III, IV respectively. To which interval would the score 7.89 be assigned? The score 14.99? The score 28.001? The score 7? The score 8? The score 15?

3. Using the method in question 2, suppose the class interval for a B was 80-89, and for an A, 90-100. Suppose you got an 89.9. Would you get an A or B for the course?

4. A histogram by construction has a total area equal to the total number of data points or one hundred percent. In the case of equal class intervals, show that the construction of the frequency polygon as described in the text guarantees that both histogram and frequency polygon have the same area, provided that a class with zero frequency is added at each end.

5. Does a frequency polygon in the case of unequal class intervals have the same area as the corresponding histogram?

6. Construct two frequency distributions from the scores of 50 students given by

56	48	48	48	44	54	46	54	50	41
46	68	46	54	51	44	40	31	46	42
55	46	58	40	43	45	43	50	26	48
54	31	38	44	52	36	35	56	56	50
43	37	52	56	26	60	48	50	54	56

Let the first distribution be one with equal intervals of 3 units each starting with 0-2, 3-5, 6-8, ... Let the second distribution be one with equal intervals of five units each starting with 0-4, 5-9, 10-14, ...
Draw the histogram and frequency curve for each distribution, and

comparing the two curves decide which of the two distributions gives a clearer presentation of the data.

7. From each of the two distributions obtained in Exercise 6, find (a) the number of students with scores between 30 and 45, (b) the number of students whose scores were at least 45. Explain why the two distributions give different answers, and find the exact answers from the original data.

8. Draw the frequency polygon and frequency curve for the distribution of the length of the left occipital bone in millimeters of old Egyptian skulls given by: Class intervals: 84-85, 86-87, 88-89, . . . , 118-119. Frequencies: 12, 12, 32, 48, 79, 116, 104, 126, 123, 74, 68, 36, 18, 7, 4, 4, 0, 1.

Chapter 4. Mean, variance, and standard deviation

"Love is a smoke made with the fume of sighs." (*Romeo and Juliet*)

Mean

The frequency distributions encountered in practice can vary considerably in their general shape. Some are symmetrical and some are skew. Some have one peak; others have two or more peaks. Ones with only one peak are called *unimodal;* ones with two peaks are called *bimodal.*

Often we wish to compare two distributions. If one is a unimodal symmetric distribution, and the other is extremely skew, then a concise comparison would be difficult to make, and so we would have to specify both distributions in detail. However often we want to compare two distributions of the same general type, and then it is possible to make a satisfactory comparison by examining only a few principal characteristics. In the case of two unimodal symmetrical distributions it is usually adequate to compare their means and variances. The *mean* is a measure of central value which essentially locates the distribution, whereas the *variance* is a measure of spread which essentially gives the degree of scatter about the mean. Often the standard deviation is used instead of the variance. The *standard deviation* is simply the (positive) square root of the variance.

There are other measures of central value such as the *median* and the *mode,* but the mean is the one most commonly used. Actually, there are three well-known kinds of means, namely the *arithmetic mean,* the *geometric mean,* and the *harmonic mean.* We are concerned with the arithmetic mean which is also called the *arithmetic average,* or simply the *average.* We want to give two formulas for computing the mean: one in the case of raw data and the other in the case of classified data (in the form of a frequency distribution). In the case of raw data the arithmetic mean is found by adding up the data and then dividing by the number of data points. For example, if the data were the scores 60, 80, 70, 85, 65 the mean is the average

$$\frac{60 + 80 + 70 + 85 + 65}{5} = \frac{360}{5} = 72$$

Let us introduce a few symbols. Let the letter x represent "any score" (i.e. any numerical observation or data point). "The sum of" is customarily denoted by the capital Greek letter sigma Σ. Then Σx means "the sum of all scores." The letter n indicates the number of scores under consideration. The average will be denoted by $\bar{x}$, which is read as x-bar. Using these symbols the formula for the mean can be expressed definitely and concisely by

$$\bar{x} = \frac{\Sigma x}{n}$$

If we had another set of n scores y then the average of these scores would be

$$\bar{y} = \frac{\Sigma y}{n}$$

In case of data in the form of a frequency distribution, the arithmetic mean is computed in the following way. Let x denote the center value of any interval, and let f denote the frequency of that interval. Then the arithmetic mean is given by

$$\bar{x} = \frac{\Sigma f x}{n}$$

where n is the total number in all the intervals, that is,

$$n = \Sigma f$$

If the frequency distribution is in terms of relative frequencies

$$f' = \frac{f}{n}$$

then the formula for the arithmetic mean becomes

$$\bar{x} = \frac{\Sigma fx}{n} = \Sigma \left(\frac{f}{n}\right) x = \Sigma f'x$$

For example, the relative frequency distribution of age for the 1960 U.S. Census given in the last chapter is:

RELATIVE FREQUENCY DISTRIBUTION OF AGE (1960 U.S. Census)

Under 5	5-19	20-44	45-64	65 and over	Total
2.5	12.5	32.5	33	Say 77.5	
0.113	0.271	0.324	0.200	0.092	100.0

The arithmetic mean is

$$\bar{x} = 0.113(2.5) + 0.271(12.5) + 0.324(32.5) + 0.200(55) \\ + 0.092(77.5)$$

which gives

$$\bar{x} = 32.3$$

We say that the average age of an American in 1960 was 32.3 years.

Variance

When central value is measured by the mean, then variability should be measured by a quantity based on deviations from the mean. Such a quantity is the variance. Let us first show how to compute the variance in the case of raw data, such as the five scores given in the preceding section. For each of the five scores 60, 80, 70, 85, 65 (with arithmetic mean $\bar{x} = 72$ as computed previously), we compute the difference of each score from the mean score; that is, we compute $x - \bar{x}$. We then do the computation shown in the following table:

Score x	Deviation $x - \bar{x}$	Squared Deviation $(x - \bar{x})^2$
60	$60 - 72 = -12$	144
80	$80 - 72 = 8$	64
70	$70 - 72 = -2$	4

85	85 − 72 = 13	169
65	65 − 72 = −7	49
Total=360	0	430

We note that the sum of the deviations from the means is always identically zero no matter how great or how small the variability; that is,

$$\sum (x - \bar{x}) = 0$$

because positive and negative deviations from the mean exactly cancel each other out. Thus, this sum provides no useful indication of variability.

As a measure of variability, we would like to have a quantity that is zero only when there is no variability (that is, when all scores are the same) and which becomes larger as the spread among scores is increased. In order that positive and negative deviations do not cancel, we first square each deviation. All of the squared deviations are positive. We then take the sum of squared deviations; for our data the sum of squared deviations is

$$\Sigma(x - \bar{x})^2 = 430$$

However, this number reflects not only the variability of the scores but also the number of scores n. As a result, we divide by $n - 1$; the result is the variance, denoted by s^2:

$$s^2 = \frac{\Sigma(x - \bar{x})^2}{n - 1}$$

(In **Chapter 8 we will give** reasons why the denominator of s^2 is $n - 1$ instead of n). For our data the variance is

$$s^2 = \frac{430}{4} = 107.5$$

For the five scores making up our data, the computation of the variance was simple because the mean $\bar{x}$ was an integer and so the deviations $x - \bar{x}$ were also integers. In general, data will not be so well behaved, and the method given will involve large rounding errors unless the deviations are carried to many decimal places. An equivalent formula for the variance which in most cases requires less arithmetic work is

$$s^2 = \frac{n\,\Sigma x^2 - (\Sigma x)^2}{n(n-1)}$$

For our data we make the table

x	x^2
60	3600
80	6400
70	4900
85	7225
65	4225
Total 360	26350

which gives

$$s^2 = \frac{5(26350) - (360)}{5(4)} = \frac{131750 - 129600}{20} = \frac{2150}{20} = 107.5$$

A third formula for the variance is

$$s^2 = \frac{\Sigma x^2 - n\bar{x}^2}{n-1}$$

which for our data gives

$$s^2 = \frac{26350 - 5(72)^2}{4} = \frac{430}{4} = 107.5$$

Let us now show how to compute the variance in the case of classified data (i.e. data in the form of a frequency distribution). The formula for the variance is

$$s^2 = \frac{\Sigma[f\,(x-\bar{x})^2]}{n-1}$$

where n is the total number in all the classes and $\bar{x}$ is the mean:

$$n = \Sigma f, \qquad \bar{x} = \frac{\Sigma f x}{n}$$

If the frequency distribution is in terms of relative frequencies

$$f' = \frac{f}{n}$$

then the formula becomes

$$s^2 = \frac{\Sigma[nf'\,(x - \bar{x})^2]}{n - 1} = \frac{n}{n - 1}\Sigma[f'\,(x - \bar{x})^2]$$

In such cases n is usually a large number so the ratio $n/(n - 1)$ is almost equal to one: hence we may then use the formula

$$s^2 = \Sigma[f'\,(x - \bar{x})^2]$$

For the relative frequency distribution of age for the 1960 U.S. Census given in the last chapter the values of $x - \bar{x}$ are −29.8, −19.8, 0.2, 22.7, 45.2 so the variance is

$$s^2 = 0.113(-29.8)^2 + 0.274(-19.8)^2 + 0.324(0.2)^2 + 0.200(22.7)^2$$
$$+ 0.092(45.2)^2$$

which is

$$s^2 = 497.622.$$

For the relative frequency distribution of age for the 1960 U.S. Census given in the last chapter the values of $x - \bar{x}$ are −29.8, −19.8, 0.2, 22.7, 45.2 so the variance is

$$s^2 = 0.113(-29.8)^2 + 0.271(-19.8)^2 + 0.324(0.2)^2 + 0.200(22.7)^2$$
$$+ 0.092(45.2)^2$$

which is

$$s^2 = 497.622$$

Standard deviation

The variance is a measure of the spread of a distribution but no graphic representation can be made of it. Its square root, however, does represent a distance that can be measured along the scale of the scores. As a matter of fact, the standard deviation acts as the standard unit in which to measure deviations of individual scores from the mean. The symbol for the standard deviation is s:

$$s = \sqrt{\text{variance}} = \sqrt{s^2} = \sqrt{\frac{\Sigma(x - \bar{x})^2}{n - 1}}$$

For the five scores 60, 80, 70, 85, 65 the standard deviation is

$$s = \sqrt{107.5} = 10.3$$

For the 1960 U.S. Census on age distribution the standard deviation is

$$s = \sqrt{497} = 22.3$$

Standard scores

Suppose that a student received a score of 83 on a mathematics examination and a score of 80 on a language examination. Can these results be interpreted that his standing is about the same on both tests? Without further information we can draw no conclusions because the raw scores do not properly reflect his relative positions on the two tests.

Suppose now that we know that the mathematics average was 75 and the language average was 65. The student stood 8 points above average in mathematics and 15 points above average in language. But can these numbers be considered comparable when the scores on the two examinations were not equally variable. If the mathematics standard deviation was 10 and the language standard deviation was 5 then the student was 0.8 standard units above average in mathematics but 3 standard units above average in language. In other words, to allow for differences in the standard deviations in different examinations, the deviation from mean in each examination is divided by the standard deviation of that examination, producing the value

$$z = \frac{x - \bar{x}}{s}$$

This z-value is called the *standard score.* In mathematics the student's standard score was

$$z = \frac{83 - 75}{10} = 0.8$$

whereas in language his standard score was

$$z = \frac{80 - 65}{5} = 3$$

Thus comparatively he did much better in language than mathematics.

Exercises

1. Two variables x and y assume the values

x:	6	-8	10	-3
y:	-8	-5	4	2

Compute

$$\Sigma x, \ \Sigma y, \ \Sigma x^2, \ \Sigma y^2, \ \Sigma xy, \ (\Sigma x)(\Sigma y), \ \Sigma(x + y), \ \Sigma x + \Sigma y$$

2. If the variable x is indeed a constant, that is, if x assumes only the constant value 5:

x:	5	5	5	5

find Σx. If the variable x is a constant with data points

x:	a	a	a	a	a	a	a	a

then what is n and what is Σx ?

3. The grades of 10 students on an exam are:

x: 55, 85, 72, 81, 83, 79, 90, 68, 85, 82

Find the mean and standard deviation. Then find the standard score for each student.

4. Prove that the sum of deviations $x - \bar{x}$ is zero. Hint: Write

$$\sum (x - \bar{x}) = \sum x - \sum \bar{x}$$

and then use Exercise 2 (as z is a constant) to find $\sum \bar{x}$.

5. If $x = y + z$ show that the average of x is equal to the sum of the averages of y and z; that is, show $\bar{x} = \bar{y} + \bar{z}$.

6. The IQ's of 500 grade school children are:

class mark	82	86	90	94	98	102	106	110	114	118
frequency	15	29	44	72	95	77	55	37	28	19

Find the mean, standard deviation, and the standard scores. Plot the frequency polygon and the frequency curve.

7. Let the x data be 4, 6, 8, 8, 4. Find $\sum x$, $\sum x^2$, $\bar{x}$, s^2. Find the z-score for each data point.

8. Which did a student do better in: A grade of 72 in a course with mean 70 and standard deviation 5 or a grade of 82 in a course with mean 78 and standard deviation 10?

Chapter 5. Events and their models

"What's in a name? that which we call a rose
By any other name would smell as sweet." (*Romeo and Juliet*)

Sets

A set is a collection of definite and distinguishable objects (or elements) selected by the means of certain rules or description. A set is represented by listing all the elements comprising it. The elements are enclosed within braces and separated by commas. A set may be defined by specifying a property that elements of the set have in common. The various objects are unordered. For example, you may have a set of six pennies. The set is the same no matter in what order you arrange the six pennies. Suppose there are 30 students in the fifth grade class. The class is the same no matter how you arrange the seating of the 30 students.

Example. The set of vowels in English alphabet is $A = \{a, e, i, o, u\}$. Since order does no matter, the set can also be written as the arrangement $A = \{e, i, o, u, a\}$ or as some other arrangement of the five vowels.

Example. The set of odd numbers less than 10 is $B = \{1,3,5,7,9\}$.

If an element x is a member of any set S, it is denoted by $x \in S$. If an element y is not a member of set S, it is denoted by $y \notin S$.

Example. f $S = \{1, 3, 5, 7\}$, then $1 \in S$ but $2 \notin S$.

A set X is a subset of set Y (written as $X \subseteq Y$) if every element of X is an element of set Y.

Example. Let $X = \{1, 2, 3, 4\}$ and $Y = \{1, 2\}$. Here set Y is a subset of set X because all the elements of set Y are in set X. Hence, we can write $Y \subseteq X$.

Let $A = \{1, 2, 3, 4\}$ and $B = \{4, 3, 2, 1\}$. Here set B is a subset of set A because all the elements of set B are in set A. Hence, we can write $B \subseteq A$. However, B is not a proper subset of A because all the elements of B are in A. If two sets contain the same elements they are said to be equal. Here A and B are equal as every element of set A is an element of set B, and also every element of set B is an element of set A.

The term "proper subset of A" can be defined as "subset of A but not equal to A." The proper subset B of set A is written as $B \subset A$. In other words, if B is a proper subset of A, then all elements of B are in A, but A contains at least one element that is not in B.

Example. Let $A = \{1,2,3,4\}$ and $B = \{1,2\}$. Then $B \subset A$.

A universal set is the set of all elements under consideration, denoted by capital U or sometimes capital E. All the sets in that context are essentially subsets of this universal set.

Example. We may define U as the set of all animals on earth. Then the set of all mammals is a subset of U, the set of all fishes is a subset of U, the set of all insects is a subset of U , and so on.

The empty set is the unique set having no elements. Its size (i.e., the count of elements in the set) is zero. The empty set is denoted by $\emptyset$. A singleton, also known as a unit set, is a set with exactly one element. For example, the set $\{0\}$ is a singleton. Two sets that have at least one common element are called overlapping sets.

Example. Let $A = \{1,2,6\}$ and $B = \{6,12,42\}$. There is a common element 6. Hence these sets are overlapping sets. Two sets A and B are called disjoint sets if they do not have even one element in common.

Example. Let $A = \{1,2,6\}$ and $B = \{7,9,14\}$. Because there is not a single common element, these two sets are disjoint sets.

Venn Diagrams

A Venn diagram, invented in 1880 by John Venn, is a schematic diagram that shows all possible logical relations between different mathematical sets. Set Operations include Set Union, Set Intersection, Set Difference, Complement of Set, and Cartesian Product.

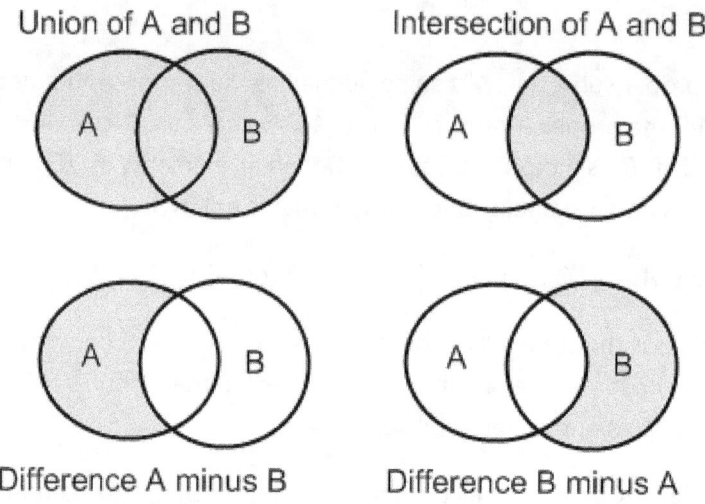

Figure 1. The set A is the oval on the left and the set B is the circle on the right. The shaded areas represents respectively: (a) the union $A \cup B$; (b) the intersection $A \cap B$; (c) the difference $A - B$; (d) the difference $B - A$.

See Figure 1. The union of sets A and B is denoted by $A \cup B$. It is the set of elements which are in A, in B, or in both A and B. The intersection of sets A and B is denoted by $A \cap B$. It is the set of elements which are in both A and B. The difference of sets A and B is denoted by $A - B$, It is the set of elements which are only in A but not in B. The difference of sets B and A is denoted by $B - A$. It is the set of elements which are only in B but not in A.

Example. If $A = \{0, 1, 2, 3, 4\}$ and $B = \{3, 4, 5\}$, then $A \cup B = \{0, 1, 2, 3, 4, 5\}$. Notice that the common elements 3 and 4 occur only once in the union.

Example. If $A = \{1, 2, 3\}$ and $B = \{3, 4, 5\}$ then $A \cap B = \{3\}$.

We recall that two sets are said to be disjoint sets if they have no element in common. Equivalently, disjoint sets are sets whose intersection is the empty set.

Example. If $A = \{0, 1, 2, 3\}$ and $B = \{3, 4, 5\}$ then $(A - B) = \{0, 1, 2\}$ and $(B - A) = \{4, 5\}$. Here, we can see that $(A - B) \neq (B - A)$

The complement of a set A is denoted by A'. It is the set of elements which are not in set A.

Example
Suppose that all red plastic toys are broken and that all red broken toys are plastic. Can you conclude that all broken plastic toys are red? We could of common sense and elementary logic to solve this problem. However, let us give a mathematical way to obtain the answer. We will use the algebra of sets. A set is a collection of certain entities (objects or elements). For the questions under discussion, it is completely immaterial what the objects are. The important thing is that the constituent elements are homogeneous. We will set up an experiment. The results (or observations) are called events. We suppose we have a group of 10 toys, of which three are red. This is an event. In fact, practically any kind of result can be termed an event. For example, any set of initial elements (called a set of elementary events) is an event.

First of all, we want introduce certain operations on events. If we have event A and event B, then it is always possible to relate two new events determined by the conditions " A and B occur" and "A or B or both A and B occur." In the former case we have an intersection $A \cap B$ of events, in the latter, a union $A \cup B$ of events. After dealing with sets for a while, you will get used to union and intersection in sets just as you are used to arithmetical operations of addition and multiplication.

We know that every horse is a plant-eating animal, but not every plant-eating animal is a horse. Let the event A be "plant-eating animal." Let the event B be "horse." Let the event C be "cow." Event A occurs every time that event B occurs. Event A occurs every time that event C occurs. A cow is also plant-eating animal. Thus event A occurs when event C occurs even though event B does not occur. For such a situation, we say that event B is included in event A or that event B is part of event A.

When an event B is included in an event A, we use the notation $B \subset A$. This is the case in Fig. 2:

$A \cup B = A$ and $A \cap B = B$

In particular,

$A \cup A = A$ and $A \cap A = A$

Fig. 2 is an illustration of this situation. Here, A and B are events consisting in a point landing in the appropriate region, and the region B lies entirely inside the region A.

In the red-toy problem, denote the set of broken toys by B, the set of plastic toys by P, the set of red toys by R. Red toys may be plastic or not plastic. In Fig. 3, the set P is hatched horizontally and the set R is hatched vertically. The set PR is cross-hatched into a grid and corresponds to red toys which are plastic.

Since it is given that all red plastic toys are broken, the set B of all broken toys must contain within it the set BR. This may be symbolized as $RP \subset B$. The situation is then depicted as shown in Fig. 84, where the set B is hatched with oblique lines. The region marked with all three kinds of hatching is the product PRB, which means plastic, red, and broken toys. The region with horizontal hatching and oblique lines corresponds to the red and broken but not plastic toys.

Such is the general situation. Now let us take into account the second assertion: "all red broken toys are plastic." Thus, there can be no red and broken but not plastic toys, which is to say the region hatched with horizontal and oblique lines must be excluded. Then, in place of the situation depicted in Fig. 84, we get Fig. 85.

This picture solves our problem completely. Namely, the region with vertical and oblique hatching indicates the possibility, in the situation described, of broken plastic but not red toys. For this reason, from the fact that all red plastic toys are broken and all red broken toys are plastic it does not follow that all broken plastic toys are red.

This can be written formally as follows: from $RP \subset B$ and $RB \subset P$ it does not follow that $BP \subset B$. Quite succinct, is it not?

All this shows that the algebra of events, a portion of which we have just explained, enables one to construct a mathematical model not only by means of ordinary elementary algebra and analysis. Incidentally, it is worth noting that Figs. 81-85 are also models of events. They are sometimes called Venn diagrams.

The algebra of events is also called Boolean algebra (after the 19th century English mathematician George Boolen or symbolic logic. The modern theory of probability rests on the Boolean algebra of events. Besides, this algebra is widely used in constructing mathematical models in many engineering problems, such as, for example, in the synthesis of relay circuits, in the theory of digital computers and the theory of finite automata.

Chapter 6. Probabiity

"O serpent heart hid with a flowering face!
Did ever a dragon keep so fair a cave?" (*Romeo and Juliet*

Random processes

One identifying feature of a random process is that it has many possible results, or cases. Since we will be speaking about the cases of a random process many times, let us give a name to the collection of all the cases. Since these cases are elemental, exclusive, and exhaustive, it is appropriate to call the set of all cases the universal set, or briefly the universe, of the random process. Another name for this universal set is the sample space.

Using this concept of universe, examples of random process may be described as follows:

Random Process	Universe (or Sample Space)
A coin is tossed.	Heads, Tails.
A die is tossed.	1, 2, 3, 4, 5, 6.
A card is drawn from a pack of 52 cards.	Ace of spades, Ace of clubs, etc., until all the 52 cards are listed.
A rain drop falls on a piece of paper.	The points on the paper.
A seed falls from a tree and lands in a meadow.	The points in the meadow.
An arrow is shot at a target.	The points on the target.

Each outcome of a random process is a happening. It is not an occurrence that can be exactly fixed in advance, perfectly predicted, etc. but instead it is an occurrence that depends in some way and to some extent upon chance, luck, hap, fortune, etc. All that we mean is that there is something about a random process that is uncertain, unsure. This uncertainty as a rule is not entirely unlimited, but only prevails in certain directions and up to certain points.

We toss a penny into the air. It happens to fall heads or it happens to fall tails. This falling is the outcome of this process of gambling. Its universe is *heads, tails,* so that the uncertainty here is limited to two cases, or three if we include the possibility of its landing on edge.

Suppose that we measure the inches of rain that falls in a year at a certain place. This quantity will lie between certain extremes which delineate the universe, say between 10 inches and 50 inches. The amount of yearly rainfall is random; that is, it depends to some extent upon unknown and unobtainable factors.

One point should be made clear. The role played by chance, luck, hap, fortune, etc., in a random process is not required to be complete but may be to any degree. For example, when a coin is tossed or a die is cast, we usually regard the role played by chance as being entire, and the role played by the person as having no effect on the ultimate result. However there may be individuals with extraordinary skill who can control some of the conditions in tossing a coin, so that they have some influence on the way it falls.

On the other hand, consider the act of shooting a bullet at a target. Because the person takes aim, the ultimate result to some extent depends upon skill. Nevertheless this act is a random process because the exact point where the bullet strikes is uncertain. The role played by the person has some effect on the ultimate result, and the role played by chance has some effect. Thus chance may be regarded as a co-agent in the final result. In fact most random processes are of the type in which chance may be regarded only as a co-agent in the ultimate result.

Statistical independence
We may classify random situations into the following two categories: (1) experimental processes, and (2) non-experimental processes. An experimental random process is one which can be repeated for an in-definite number of instances. In each instance the same circumstances are fulfilled and one result is produced. Thus if the experimental process

is repeated five times, a series of five results is produced. All other processes fall into the category of non-experimental processes.

Experimental processes are very useful in gaining insight to the methods of probability theory. For brevity, we shall replace the long expression experimental, random process by the single word *experiment*. Thus an experiment is a random process that can be repeated indefinitely, where each repetition is made under the same circumstances and produces one result. Twenty repetitions of the experiment would produce a series of twenty results. Each result, of course, must be one and only one of the possible results of the experiment. An example of an experiment is the tossing of a coin. This experiment can be repeated an indefinite number of times. For each repetition, either H or T happens. Thus a series produced by 20 repetitions might be

$$HHHTTHHTHTHHTTTHTHTT$$

The intuitive concept of experiments repeated under identical conditions leads to the notion of *statistical independence.* When a scientist says that two experiments are performed under identical conditions, he implies independence, that is, he implies that the result of either experiment has no influence on the result of the other experiment. For example, if we toss a coin twice under identical conditions, then the two tosses are independent. The appearance of heads or tails on toss 1 does not influence which side appears on toss 2.

The notion of statistical independence applies not only to repetitions of the same experiment under identical conditions, such as successive tossing of a coin, but also applies to different experiments that have nothing to do with each other. Such experiments can either be done simultaneously (i.e. at the same time) or sequentially (i.e. in a time succession). For example, suppose we toss a coin and throw a die, either simultaneously or sequentially. It seems evident that the fall of the coin has nothing to do with the fall of the die, and accordingly we say that these two experiments are independent.

Classic definition of probability

Now we want to look at random processes that have a finite number of possible results. Thus we may list all the cases for such a situation. In tossing a coin there are two cases; namely H, T. In tossing a die there are six cases, namely 1, 2, 3, 4, 5, 6. For many such finite processes, there exists a symmetry among the cases. For example, a die is in the form of a cube and is manufactured from a homogenous material. Thus its six faces have symmetry that tells us that the turning up of one side should not be more propitious than the turning up of any other side. We must use our judgement to decide for each particular process whether its cases do or do not have the necessary kind of symmetry. In other words, for the process in question we must decide whether or not any case is more propitiously endowed than any other case. To make this decision we must make use of our accumulated experience as well as theoretical reasoning by analogy.

When the cases of a random process do have this kind of symmetry, that is, when no case is more propitious than any other case, then each case is *equally likely*. An *equally-likely case is* called a *chance.* We may then define the *probability* of an event as the ratio of the number of favorable chances to the total number of chances. In symbols we write *P(A)* to stand for the probability of event A. Then the definition is

$$P(A) = \frac{\text{number of favorable chances}}{\text{total number of chances}}$$

This definition is called the classic definition of probability. For example, when we toss a good coin, H has 1 chance in 2, so the probability of H is 1/2. When we throw a good die, the event 5 (i.e. the face 5 lands up) has 1 chance in 6, so its probability is 1/6. When we say that the event of each face on a good die has a probability of 1/6, we always must take into consideration that a so-called "good" die can only be realized in practice to a certain approximation. The same situation holds for all the questions studied by means of probability theory.

The difference between a mathematical model and a mathematical theory should be noted. A mathematical theory does not need any connection whatsoever with the real world. A mathematical model is a mathematical theory that can be applied to real phenomena. The definition of probability given here represents a mathematical model. There are many applications of this model to real phenomena. In fact, this probability model was discovered in the Renaissance by the empirical study of some real phenomena, namely the tossing of dice and the dealing of cards in games of chance.

Relative frequency definition of probability

Let us now see how we can interpret a probability, as say the probability of 1/6 that the face 5 lands up in the toss of a die. We make use of the so-called *Law of Large Numbers,* which states:

> If an experiment is repeated more and more times, then the relative frequency of a given event tends closer and closer to the probability of the event.

If the die is cast again and again, then the relative frequency

$$\frac{\text{number of times face 5 lands up}}{\text{total number of tosses}}$$

will fluctuate, but it comes closer and closer to the value 1/6, namely the probability of the face 5 landing up. As a result we can interpret the probability of an event as the proportion of the times it will occur in the long run. In saying that the probability is 1/6, we mean that if the experiment is repeated a great many times the event will occur about 16 2/3 percent of the time. Note that we do not say that the event must occur 1 time out of 6, or 10 times out of 60, or 100 times out of 600. We only say that the relative frequency of the event in 600 tries will generally be closer to the true probability 1/6 then the relative frequency of the event in 60 tries. In turn, the relative frequency in 60 tries will be a better estimate of the probability 1/6 than the relative frequency in only 6 times. In short, things take time.

As a consequence of the statistical regularity indicated by the law of large numbers we now have a way of estimating the values of probabilities. Many events cannot be put into the framework of the classical definition of probability, which requires the existence of equally likely cases. For example, let us observe a number of automobiles of a given year and model, and suppose we want to predict whether each automobile five year hence is still in operation. Let us suppose that with expert mechanics we are able to make detailed examinations of each car, and also are able to collect pertinent information as to the occupation and habits of the owners. However it is impossible to make exact predictions with regard to the outcome of one particular automobile, since the causes leading to the ultimate result are far too numerous and too complicated to allow any precise calculation. We can think of the five year time-span of each car as an experiment, and the experiment is a success if the car is still in operation after five years. As we have said we are not able to predict individual results, but as soon as we turn our attention from the individual cars to the whole sequence of cars we find that the relative frequency of successes shows a striking regularity as we include more and more cars. More specifically, the relative frequency for more and more cars shows a marked tendency to become more and more constant. If such a series of cars could be indefinitely continued under uniform conditions, then we would expect the relative frequency to approach some definite ideal value which we call the probability. It is a common experience to mankind that this stability of a relative frequency usually appears in long series of repeated experiments performed under uniform conditions. Moreover in situations where this statement is not true, a careful examination will usually disclose some definite lack of uniformity. However this statement represents a conjecture that can neither be proved nor disproved either mathematically or empirically, and all we can say is that its validity is supported by our experiences in life. The resulting meaning assigned to the word *probability* is called the *frequency interpretation of probability.*

We can, therefore, define the probability of an event as the limiting ratio of the number of favorable results to the total number of results, as the total number of repetitions of the random experiment becomes larger and larger. The definition can be written as

$$P(A) = \frac{\text{number of favorable results}}{\text{total number of results}}$$

where it is understood that the total number of repetitions gets very large. Because this ratio is the relative frequency of the event A, we call this definition the *relative frequency definition of probability.*

In real life there is an upper bound to the number of times we can perform an experiment, so we can never be sure that we have obtained a large enough total to insure the accuracy of our probability estimate. In such cases we must do the best we can. For example, if we could only study the history of 1000 automobiles, and found that 650 were still in operation after 5 years, we would assign the relative frequency 650/1000 = 0.65 as the probability of success.

Personal, or subjective, definition of probability

There is a third definition of probability, which is often useful in the case of a single non-repetitive event. The *personal, or subjective, definition of probability* says that the probability is the measure of a person's belief as to the occurrence of the event. Such personal probabilities are arrived at by subjective reasoning that is difficult or impossible to pin down to a formula or method as we were able to do in the case of the other two definitions. However the concept is useful, especially in dealing with situations where there is little or no direct evidence. Sometimes such probabilities are referred to as intuition or guesses. For example, consider the probability of success in a new venture in which neither a scientist nor anyone else has had any previous experience. The scientist subjectively might obtain the figure 0.75, or 3 chances out of 4. As a matter of fact, people often assign a probability to an event by means of insight or some hunch.

Chapter 7. Events

"There's an old saying that applies to me:
you can't lose a game if you don't play the game." (*Romeo and Juliet*)

Definition of event

As we have seen the universe of a random process is made up of all the possible results, or cases. These cases are elemental (i.e., part of a case is not admitted), exhaustive (i.e., one case happens) and exclusive (i.e., only one case happens). In brief, whenever the circumstances of a random process are fulfilled, one and only one case happens. For example, in the tossing of a die there are six cases: 1, 2, 3, 4, 5, 6. An event is a set of cases. In our die example, the event odd is the set consisting of the three cases 1, 3, 5. The cases that make up an event are called the members of the event. We say that an event happens when and only when one of its members happens.

The *universal event* is the event made up of all the cases. It is the universe of the random process under consideration. Because the universal event always happens, it is a sure event and we assign it a probability of one. The probabilities of all the events must lie between 0 and 1.

The *empty* (or *null) event* is the event made up of none of the cases. In mathematics, and more specifically set theory, the empty set is the unique set having no elements; i.e., its count of elements is zero. The empty set is designated by the symbol $\emptyset$. A null event is an event that is impossible. More precisely, since an event is a subset of a sample space, the null event is the empty set. Because the null event never happens, we assign it a probability of zero; i.e., $P(\emptyset) = 0$. However, in some situations, a nonempty event may have probability 0.

The complement (or contrary) of any event A is the event "not A"; i.e., the event that A does not occur. The complement of event A is usually denoted as A^c, A' or $\bar{A}$. We will use A^c. The event A and its complement A^c are mutually exclusive and exhaustive. If we remove the members of A from the universe, then we are left with the complement event A^c. The two events A and A^c have no members in common, but everything in the universe is either a member of A or a member of A^c. For example, in the tossing of a die, the event *odd* is made up of the number 1, 3, 5. The complement event A is thus made up of 2, 4, 6, and is in fact the event *even*. If the probability of the event A is $P(A)$ then the probability of the complement event is

$$P(A^c) = 1 - P(A)$$

In other words, the sum of the probabilities of an event and the complement event is equal to one.

Two events can have members in common. For example, in the toss of a die, the event *low* is made up of the cases 1, 2, 3, whereas the event *odd* is made up of the cases 1, 3, 5. Hence these two events have the cases 1 and 3 in common. If either 1 or 3 occurs, then both the event *low* and the event *odd* happens.

Two events can have members in common. For example, in the toss of a die, the event *low* is made up of the cases 1, 2, 3, whereas the event *odd is* made up of the cases 1, 3, 5. Hence these two events have the cases 1 and 3 in common. If either 1 or 3 occurs, then both the event *low* and the event *odd* happens. Let us designate the event made up of the cases in common, namely 1 and 3, by the notation

low ∩ even

which we read as *low intersection even.* That is, the *common event* $A \cap B$ is defined as the event made up of the common members of the separate events A and B. These common members are cases that are members of both of the events A and B.

Two events are *incompatible* provided that they have no members in common. In other words, two events are incompatible if and only if they cannot both happen at the same time. Alternate names for incompatible events are *disjoint events* or *mutually exclusive events*. Because the common event $A \cap B$ is made up of the members common to the events A and B, we see that the events A and B are incompatible if and only if their common event $A \cap B$ is the empty set. The empty set is designated by the symbol $\emptyset$. A null event is an event that is impossible. More precisely, since an event. For example, the events *odd* and *even* are incompatible. An event and its complement event are always incompatible, that is,

$$A \cap A^c = \emptyset$$

People who swim and play tennis can form a union. The union is made up of people who swim or play tennis. But suppose a person both swims and plays tennis. Certainly he will be a member of the union, but he will be counted as only one member, not as two members. Thus he has no preferential status to a person who swims but does not play tennis, who is also counted as one member of the union. Likewise, a person who plays tennis but doesn't swim is counted as one member. Suppose there are 100 people who swim (regardless of whether or not they play tennis), 70 people who play tennis (regardless of whether or not they swim) and 20 people who both swim and play tennis. Thus the union has

$$100 + 70 - 20 = 150 \text{ members}$$

In general, given any two events A and B we can form their *union*

$$A \cup B$$

which we read as A *union* B. The event $A \cup B$ is the event whose cases are members of at least one of the events A and B.

When the events A and B are incompatible (i.e. A and B have no members in common) then the members of the union $A \cup B$ are simply all the members of A together with all the members of B.

Because there is no member that both belongs to A and B, there is no possibility of double-counting such a member.

In other words, the sum of the probabilities of an event and the

Addition rule of probability for incompatible events

One of the most important, and also the simplest, rules used in the calculation of probabilities is the *addition rule in the case of incompatible events*. For example, on Nevada roulette there are 18 chances in 38 for red and 18 chances in 38 for black. Let A stand for red, and B for black. Note the 0 and 00 on the Nevada roulette wheel are neither red nor black. The events red, black are incompatible. Either red can happen or black can happen, but not both. Thus the united event $A \cup B$ has 18 + 18 = 36 chances in 38 to happen. Thus the sum of the probability that red happens plus the probability that B happens is

$$P(A \cup B) = P(A) + P(B) = \frac{18}{38} + \frac{18}{38} = \frac{36}{38}$$

We may summarize our reasoning as follows:

> We first verify that two events A, B are incompatible; that is $A \cap B = 0$. Then the probability of their united event $A \cup B$ is obtained by adding their separate probabilities; that is,

$$P(A \cup B) = P(A) + P(B)$$

The addition rule in the case of incompatible events is treated in the next chapter.

Expectation of an event

We now come to an important topic; namely, expectation. The probability of an event is not the only quantity in which we are interested in the study of a random process. For example, it is very different to have 1 chance in 10 to gain $100 or to have 1 chance in 10 to gain $100,000. This consideration leads to the notion of *mathematical expectation*, or to use a shorter term *expectation*, of an event.

If a person has 1 chance in 10 to gain $100, we say that his expectation is

$$\$100\left(\frac{1}{10}\right) = \$10$$

If he has 1 chance in 10 to gain $100,000, his expectation is

$$\$100,000\left(\frac{1}{10}\right) = \$10,000$$

If a person has 1 chance in 2 to gain $200, his expectation is

$$\$200\left(\frac{1}{2}\right) = \$100$$

Suppose that the only prize given at a lottery is $100,000, and that there are 200,000 tickets sold at $1 each. The expectation of a ticket is then

$$\$100,000\left(\frac{1}{200,000}\right) = \$0.50$$

even though the cost of the ticket is $1. Examining this situation, we see that the people who organized this lottery realize $200,000 from ticket sales, and only pay out $100,000 as a prize, thereby making a profit of $100,000. On the other hand each of the 200,000 ticket holders paid $1 for a ticket whose expectation was only $0.50.

The *expectation* of a contingent gain is *the product of the gain times the probability of realizing this gain.* We can say that the expectation is the value of a gain whose attainment is not certain.

When it is a matter of a loss instead of a gain, we shall consider a loss to be a negative gain. For example, if a person has 1 chance in 50 to lose $100, his expectation is

$$\$100 \left(\frac{1}{50}\right) = -\$2$$

The expectation is negative. A negative expectation is the value of a loss whose realization is not certain but only contingent. If there is a probability of 1/20 to sustain a loss of $800, the expectation is

$$-\$800 \left(\frac{1}{20}\right) = -\$40$$

In order to avoid this risk of losing $800, we may say that $40 is the "fair" amount that should be paid.

Suppose that for a toss of a coin, a player receives $100 if the coin lands H (whereas before H stands for heads). The probability of H is 0.5. We say that the expectation of the player is

$$\$100(0.5) = \$50$$

Suppose that for two consecutive tosses of a coin, another player receives $100 if both tosses result in H. The probability that both tosses land H is 0.25. The expectation of the player is then

$$\$100(0.25) = \$25$$

The expectation of a player is the product of his possible gain times the probability of his realization of this possible gain.

Summing up, we have: *The expectation of an event is the product of the gain received if the event happens times the probability that the event happens.*

The expectation is a fictitious or imaginary sum of money. It does not ordinarily correspond to a possible value of gain or of loss. If a person has 1 chance in 10 to gain $40 his expectation is $4. But $4 is not a possible value of his gain. The only possible values that can happen are $0 and $40. If a person has 1 chance in 5 to lose $10, his expectation is a loss of $2. Nevertheless, either a loss of $0 or a loss of $10 will happen, never a loss of $2.

Expectation of several events

An advantage of the notion of expectation is the following. The combination of different probabilities can be complicated, but the combination of different expectations is simple and intuitive. It is easy to see that expectations can be added together as sums of ordinary money.The rule is:

The expectation for several events is the sum of the expectations for each of the events.

This property of addition can make the calculation of expectations very easy in many applications. Suppose a coin is tossed 3 times. If it lands H on toss 1 a player receives $10. If it lands H on toss 2 he receives $20. If it lands H on toss 3 he receives $30. Thus we have

Event	Gain	Probability	Expectation
H on toss 1	$10	0.5	$5
H on toss 2	$20	0.5	$10
H on toss 3	$30	0.5	$15

Therefore his total expectation is $5+$10+$15=$30

As we have seen, a negative gain is a loss. In dealing with expectations, we must be prepared to consider negative gains (or losses) as well as positive gains.

Suppose a coin is tossed. A player receives $100 if the coin lands H and loses $100 if the coin lands T. What is his expectation? The loss of $100 is the same as a gain of −$100. Hence we have the table:

Event	Gain	Probability	Expectation
H	$100	0.5	$50
T	−$100	0.5	−$50

Hence his total expectation is $50 − $50 = 0.

Advantageous, equitable, and disadvantageous games

A game is *advantageous* to a player if his expectation is positive. It is *disadvantageous* if his expectation is negative. It is neither advantageous nor disadvantageous when his expectation is zero. We then say that the game is *equitable.*

For example, suppose a player has 1 chance in 3 to gain $24 and 2 in 3 to lose $12. The game is equitable because his expectation is

$$\$24\left(\frac{1}{3}\right) - \$12\left(\frac{2}{3}\right) = 0$$

If a game is made up of several trials, the expectation of the game is the sum of the expectations of the various trials that compose the game. It is understood that all the trials are necessarily played. In particular, if all the trials are identical, the expectation of the game (composed of n trials) is equal to the product of n times the expectation of one trial. Thus if each trial is equitable, the game is equitable. As a result there is no way to make advantageous or disadvantageous a game of which each trial is equitable.

Betting on events

Gambling is the act of betting on unsure events. Life and people's ingenuity provide ample events upon which people can bet. A bettor is a person who bets, typically regularly or habitually. To place a bet one must first know the odds given for the event. The *bettor's odds* consist of two numbers separated by the word *to,* such as 8 to 5. The bettor's odds represent the ratio of his potential gain to his potential loss; that is, odds = gain to loss. For example, suppose his odds are 8 to 5 for the event, and the bettor bets $5 on the event. If the event happens he wins. His winnings are $13, made up of his $8 gain plus his $5 bet. If the event doesn't happen he loses. His loss is $5, made up of his $5 bet.

Another way of thinking about the bettor's odds is that the odds represent *his potential gain for his bet.* For example, suppose the odds are 6 for 1 for an event, and a bettor bets $10 on the event. If the event happens, his gain is $60 for his $10 bet, making his total purse equal to $70. If the event does not happen, his loss is his bet of $10.

Application to life insurance

According to the mortality tables of an insurance company, the probability that a 25 year old man will live at least one more year is 0.993, whereas the probability that he will die within the year is 0.007. A $10000 1-year term life insurance policy is an agreement whereby the insurance company will pay to the man's survivors $10000 if the man dies within the year, but pays nothing if the man lives for the entire year. The company offers such a policy to a 25-year old man for a premium (that is, cost to the man) of $100.

The company faces the following situation:

Event	Gain to Company	Probability	Expectation
Man lives	$100	0.993	$99.30
Man dies	−$9900	0.007	−$69.30

Hence the expectation of the company is $99.30 minus $69.30, which is $30.00 from which the company must pay costs, taxes, and dividends. If this expectation were negative, the company could not stay in business. As a game, life insurance is advantageous to the company and disadvantageous to the policy holder.

Exercises

1. Games may be divided into three categories: (1) pure chance, (2) chance and skill, (3) pure skill. Into which categories would you place chess, dominos, most card games, roulette, dice? It is generally impossible for a card player to have in his mind all of the different probabilities. There is a prodigious number of different possible results that can be obtained from a pack of cards. The number of possible arrangements of 52 cards is expressed by an 8 followed by 67 other digits. Since the player is not able to completely know all the probabilities, can he appeal to his experience and intuition to

supplement his partial knowledge, so that his skill and cleverness play a part, and the game no longer appears as completely dependent upon chance?

2. In a game of coin tossing, dice, or roulette a player can have in his mind the probabilities of the different events which he is faced with. Can we say that a knowledgeable player is one who knows these probabilities and acts on their basis, whereas an unknowledgeable player is one who does not know the probabilities and is under the illusion that his skill and cleverness play a part?

3. Player A has 3 silver dollars and B has 2 silver dollars. The coins are all tossed, under the agreement that the player having the greatest number of heads shall win all the 5 silver dollars, but in case of a tie B shall win. What is the expectation of A?   What is the expectation of player A if the rules are changed so that the 2 players agree to begin again in case of a tie?

4. The French roulette wheel is used in Monte Carlo. It has 37 numbers: 0, 1, 2, 3, 4, . . . , 34, 35, 36. The casino pays odds of 35 to 1 for single-number events. What is the expected loss of a bet of 100 francs played on a single number? *Ans.* 2.70

5. The event *odd* on the French roulette wheel is made up of the 18 numbers: 1, 3, 5, . . . , 33, 35, whereas the event *even* is made up of the 18 numbers: 2, 4, 6, . . . , 34, 36. If you bet 100 francs on *odd* then (1) if *odd* happens, your winnings are a gain of 100 francs plus your bet of 100 francs, (2) if *even* happens, you lose your 100 francs bet, and (3) if 0 happens, your bet is left where it stands and the wheel spun again. On this second spin: (1) If *odd* happens you take your bet of 100 francs back, (2) if *even* happens you lose your 100 francs, and (3) if 0 happens, the wheel is spun a third time, and so on. What is your expected loss? *Ans.* 1.35

6. Two dice are thrown. What is the probability that they show 7 or 11?

Chapter 8. Rules for probabilities

"Love is heavy and light, bright and dark, hot and cold, sick and healthy, asleep and awake- it's everything except what it is! This is the love I feel, though no one loves me back." (*Romeo and Juliet*)

Addition rule for probabilities

The fundamental rule upon which the rule for the addition of probabilities rests is the addition rule. The *addition rule* is:

> If one thing can be done in m different ways and another thing can be done in n *other* different ways, then either one thing or the other can be done in the sum of m plus n different ways.

This rule is basic, and goes back to when you first learned how to add. For example, if we put 5 apples and 6 oranges into a box, then the number of apples and oranges in the box is 5+6, or 11.

As another example, if John knows 4 kinds of trees and Mary knows 3 other kinds, then John or Mary know 4+3=7 different kinds. Suppose John knows oak, elm, maple and birch, and Mary knows cedar, pine, and spruce. Then oak, elm, maple, birch, cedar, pine and spruce are known by John or Mary. However, if John knows 4 kinds of trees, and if Jane knows 3 kinds, then we cannot conclude that John or Jane know 4 + 3 or 7 different kinds. The reason is that Jane might not know 3 *other* kinds, but instead knows some of the same kinds as John. Suppose that Jane knows oak, elm and pine. Then oak, elm, maple, birch, and pine are known by John or Jane, which are 5 different kinds.

Thus the word *other* is an essential part of the addition rule.

A set that has only a finite number of members is called a finite set. Let this finite number be denoted by n. The number n is a positive whole number, such as 1, or 2, or 5, or 5000, or 500,000. We exclude $n = 0$, which would correspond to the empty set. Instead of saying "a set with n numbers," it is much more convenient to use a shorter expression. Hence let us agree that the expression "n-set" means "a set with n members."

Now we can state the *addition rule* in terms of sets.

As we know, the symbol Ø denotes the empty set. Let A be a m-set and B be a n-set. Let A and B have no members in common; that is, let the common set satisfy

$$A \cap B = \emptyset$$

Then the united set

$$A \cup B$$

is an $(m + n)$-set.

Let us now make use of the classic definition of probability. We deal with a random process that has these two properties:

   (1) The process has only a finite number of cases.
   (2) No case is more propitiously endowed than any other case.

The *classic definition* is:

   The probability of an event is the ratio of the number of favorable cases to the total number of cases.

Suppose we have an urn that holds 20 like balls, among which:

   5 are colored amber *(A)*
   7 are colored blue *(B)*
   8 are other colors.

Let the desired event be the event of extracting an amber or blue ball. This event is denoted by

$$A \cup B$$

There are 2 kinds of favorable cases for this event. One kind is represented by the amber balls, that is, the cases of the event A. The other kind is represented by the blue balls; that is, the cases of the event B. The event A has no cases in common with the event B, because if the extracted ball is amber then it cannot be blue, and if the extracted ball is blue then it cannot be amber. Thus the two events A, B are incompatible; that $is\, A \cap B = \emptyset$. The event A has 5 cases in 20. The event B has 7 cases in 20. The 5 cases of A are distinct from the 7 cases of B. These two events cannot both happen at the same time. Because the events A, B have no cases in common, the number of cases for the event $A \cup B$ is the sum of the cases for A plus the cases for B. That is, the number of cases for the event $A \cup B$ is Because the cases add, the probabilities also add. Hence the probability of the event $A \cup B$ is the sum of the probability of A plus the probability of B. That is,

$$P(A \cup B) = \frac{5}{20} + \frac{7}{20} = \frac{12}{20}$$

This equation illustrates the rule for the addition of probabilities:

$P(A) + P(B) = P(A \cup B)$ provided $A \cap B = \emptyset$.

For ease of memory, it may be stated as follows:

> If the events A, B are incompatible, then the probability of the united event $A \cup B$ is equal to the sum of the probability of A plus the probability of B.

The rule states, simply, that the probabilities of two incompatible events can be added. The sum of the probabilities is the probability that one or the other of these two incompatible events happens.

In the application of this rule, it is very important to verify that the condition of incompatibility is fulfilled. Suppose that there are 5 air routes from London to New York and 3 air routes from London to Moscow. These sets of air routes are incompatible. Thus the number of air routes going from London to either New York or Moscow is

$5 + 3 = 8$

Suppose that there are 30 planes at the London airport, and each plane is going to fly a different air route. You pick one airplane at random and get it. Then you have 5 chances in 30 to go to New York, and 3 chances in 30 to go to Moscow. Thus you have 8 chances in 30 to go to either New York or Moscow, and so the probability of this event is 8/30.

Now let us suppose that 1 of the 5 air routes from London to New York includes a stop at Iceland. Also suppose there is no other air route from London to Iceland. Although there are 5 air routes to New York and 1 air route to Iceland, the number of air routes to either New York or Iceland is not 6, but 5. In other words, the air route to Iceland is the same as one of the air routes to New York, and so the event of going to Iceland is not incompatible with the event of going to New York. Your probability of going to Iceland or New York is 5/30, not 6/30.

Problem: Three coins are tossed. What is the probability of "at least $1T$"? Note: T=tails and H=heads.

Solution: The tossing of 3 coins has 8 cases in total; namely, the eight cases are

$$HHH, HHT, HTH, THH, HTT, THT, TTH, TTT$$

The event $1T$ (i.e. 1 tails shows) has 3 cases, namely

$$HHT, HTH, THH$$

The event $2T$ (i.e. 2 tails show) has 3 cases, namely

$$HTT, THT, TTH$$

The event $3T$ (i.e. 3 tails show) has 1 case, namely

$$TTT$$

The event of "at least $1T$" is the union of the events $1T$, $2T$, $3T$; that is,

$$\text{at least } 1T = 1T \cup 2T \cup 3T$$

The events $1T$, $2T$, $3T$ are incompatible. Hence the event of "at least $1T$" has $3 + 3 + 1 = 7$ chances, so its probability is 7/8. Because

$$P(1T) = \frac{3}{8}, \qquad P(2T) = \frac{3}{6}, \qquad P(3T) = \frac{1}{8}$$

we see that

$$\frac{7}{8} = \frac{3}{8} + \frac{3}{8} + \frac{1}{8}$$

or

$$P(1T \cup 2T \cup 3T) = P(1T) + P(2T) + P(3T)$$

The equation represents the addition rule for probability for three incompatible events.

Problem: Ann and Bob toss a coin under the following conditions. If toss 1 shows H, Ann wins. If toss 1 shows T, however, the coin must be tossed 2 more times. Then, if out of the 3 tosses, H shows at least 2 times, Ann also wins. What is the probability that Ann wins?

Solution: We could reason as follows. Ann wins in 2 different ways. The first way is for H to come up on toss 1. The probability of this event is 1/2. The second way is for H to come up at least 2 times out of 3 tosses. The probability of this event is also 1/2. Thus the probability that Ann wins is the sum of these 2 probabilities. This sum is

$$\frac{1}{2} + \frac{1}{2} = 1$$

which says that it is certain that Ann wins. But this is absurd. We must consider the set of all possible outcomes; namely,

$$HHH, HHT, HTH, THH, HTT, THT, TTH, TTT$$

Now the cases favorable to Ann winning by the first way (namely, by H coming up on toss 1) are

$$HHH, HHT, HTH, HTT$$

so indeed the probability is 4/8, or 1/2, for her winning by the first way. Also the cases favorable to Ann winning by the second way (namely, by H coming up in 2 out of three tosses) are

$$HHH, HHT, HTH, THH$$

so the probability is 4/8, or 1/2, for her winning by the second way. But for 3 of these cases, namely

$$HHH, HHT, HTH$$

Ann wins directly by the first way (namely, H on toss 1), and so toss 2 and 3 need not be made. Thus the two ways that Ann might win are not incompatible, so it was not legitimate to add their probabilities. In other words, the chances

$$HHH, HHT, HTH$$

are common to both ways. The chances favorable to Ann winning one way or the other are

first way

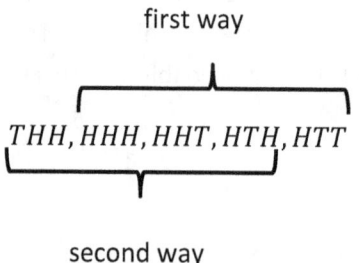

$$THH, HHH, HHT, HTH, HTT$$

second way

so her chances are 5 in 8. Thus the probability of Ann winning is 5/8.

Multiplication rule for probabilities

Another general rule, which like the addition rule is extremely useful, is the *multiplication rule.* It is

> If one thing can be done in M different ways, and then another thing can be done in N different ways, it follows that both things can be done in the product $M \times N$ different ways.

Another statement of the *multiplication rule* is

> If there is an M-way choice for thing 1 and *then* there is an N-way choice for thing 2, it follows that there is an $M \times N$-way choice for both things.

This rule is basic. It was first taught when you learned how to multiply numbers. For example, if you first choose one of coffee, tea, or milk and then either a ham sandwich or a peanut butter sandwich, there are $3 \times 2 = 6$ different choices that you have; namely,

coffee, ham sandwich	coffee, peanut butter sandwich
tea, ham sandwich	tea, peanut butter sandwich
milk, ham sandwich	milk, peanut butter sandwich

Suppose there are 3 trials to the top of Mount Washington. Thus there are 3 ways up and 3 ways down, so that there are $3 \times 3 = 9$ ways of making one round trip. If the trials are called A, B, C then these 9 round trips are

AA	AB	AC
BA	BB	BC
CA	CB	CC

Suppose now that we do not want to come down by the same trial as the one we went up. Now there are 3 trials to go up, but only 2 trials

that we can come down on. Hence the number of round trips is
$3 \times 2 = 6$. They are

AA	AC
BA	BC
CA	CB

Problem 1: Two persons get into an airplane where there are 9 vacant seats. In how many different ways can they seat themselves?

Solution 1: Person 1 can take any of the 9 vacant seats. Thereafter person 2 can take any of the 8 seats that are left. Hence there are $9 \times 8 = 72$ different ways that they can take their seats.

The multiplication rule is very useful in computing probabilities. In the problems here we make use of the fundamental classic definition:

$$\text{probability} = \frac{\text{number of favorable cases}}{\text{number of all cases}}$$

Problem 2: Two cards are drawn from a well-shuffled pack of 52 cards. What is the probability that both cards drawn are kings?

Solution 2: Since there are 52 cards in the pack, the first card can be drawn 52 ways. After the first card has been withdrawn, there are 51 cards remaining in the pack, so the second card can be drawn in 51 ways. Therefore, the total number of ways to draw 2 cards is $52 \times 51 = 2652$, which is the number of all the cases. To find the number of cases favorable to the event of drawing two kings, we observe that there are 4 kings in the pack. Thus the first king can be drawn in 4 ways. After the first king has been drawn, there are 3 kings left, so the second king can be drawn in 3 ways. Therefore the number of favorable cases is $4 \times 3 = 12$. Hence there are 12 cases in 2652 that both cards drawn are kings, so the required probability is

$$P(2K) = \frac{4 \times 3}{52 \times 51} = \frac{1}{13 \times 17} = \frac{1}{221}$$

That is, there is 1 chance in 221 of drawing 2 kings if the first card is not returned to the pack before drawing the second card.

Problem 3: Two cards are drawn from a pack of 52 cards, the first card being returned to the pack before the second card is drawn. (By this description we mean: The cards are shuffled and the first card is drawn and noted. This card is then returned to the pack, the pack is shuffled again, and the second card is drawn and noted.) What is the probability that both cards drawn are kings?

Solution 3: There are 52 ways of drawing the first card. There are also 52 ways of drawing the second card, because by returning the first card drawn, the pack is restored to its original number of 52 cards. Thus the number of ways to draw both cards is 52×52. There are 4 ways of drawing the first king, and because the pack is restored, there are also 4 ways of drawing the second king. Thus the number of ways to draw both kings is 4×4. Hence the required probability is

$$P(2K) = \frac{4 \times 4}{52 \times 52} = \frac{1}{13 \times 13} = \frac{1}{169}$$

That is, there is 1 chance in 169 of drawing 2 kings if the first card drawn is returned to the pack before drawing the second card.

Problem 4: 3 chests, identical in appearance, each have 2 drawers. Chest 1 has a G (Gold) coin in each of drawers 1 and 2. Chest 2 has a S (Silver) coin in each of drawers 1 and 2. Chest 3 has a S (Silver) coin in drawer 1 and a G (Gold) coin in drawer 2. See the table below

	Chest 1:	Chest 2:	Chest 3:
Drawer 1	G	S	S
Drawer 2	G	S	G

A chest is chosen at random. What is the probability that its 2 coins are of different metals.

Solution 4: Since outwardly the chests are indistinguishable from each other, we recognize each chest as having 1 chance to be drawn. Among the 3 chests, only 1 chest has coins of different metals, namely chest 3. Therefore the required probability is 1/3.

Problem 5: Let there be 3 chests as in the foregoing problem. A chest is chosen at random, one of its drawers is opened, and a S (silver) coin is found. What is the probability that the other drawer contains a G (gold) coin?

Solution 5: The fact that S was found in one drawer leaves only 2 possibilities as to the content of the other drawer, namely S, G. Hence we might be tempted to reason that the probability of G in the other drawer is 1/2. Nevertheless, this reasoning is false, because each of the possibilities cannot be regarded as having 1 chance each. Before the chest is chosen and one drawer opened, there are 6 possible results, or cases; namely,

Drawer 1 of chest 1	Drawer 1 of chest 2	Drawer 1 of chest 3
Drawer 2 of chest 1	Drawer 2 of chest 2	Drawer 2 of chest 3

Each one of the 6 represents 1 chance. But as soon as S is found in the drawer that is opened, the 3 cases

Drawer 1 of chest 1		
Drawer 2 of chest 1		Drawer 2 of chest 3

became impossible, and so there remain only 3 cases, namely

	Drawer 1 of chest 2	Drawer 1 of chest 3
	Drawer 2 of chest 2	

Each of these cases represents 1 chance. In these 3 chances, there is 1 chance, namely Drawer 1 of chest 3, that the other drawer contains G. There the required probability is 1/3.

We may extend the *multiplication rule* to doing 3 things, one after another. Thus, we have:

If thing 1 can be done in N_1 different ways, and if *then* thing 2 can be done in N_2 different ways, and if *then* thing 3 can be done in N_3 different ways, it follows that all the things can be done in $N_1 N_2 N_3$ different ways.

Problem 6: 10 persons compete for 3 prizes. In how many different ways can the prizes be awarded?

Solution 6: Prize 1 can be given in 10 different ways. Thereafter prize 2 can be given in 9 different ways. Thereafter prize 3 can be given in 8 different ways. Hence there are (10) (9) (8) = 720 different ways to give the prizes.

The *multiplication rule* can be extended to doing any number of things.

If series of things can be done successively as follows:
Thing 1 in N_1 different ways,
then thing 2 in N_2 different ways,
then thing 3 in N_3 different ways, $\cdots$
then thing k in N_k different ways,
then all the things can be done in $N_1 N_2 N_3 \cdots N_k$ different ways.

Problem 7: The probability that a newspaper reader will read an advertisement is 0.3 and if he reads the advertisement the probability that he will buy the product advertised is 0.02. What is the probability that he will read the advertisement and then buy the product?

Solution 7: The tree diagram is:

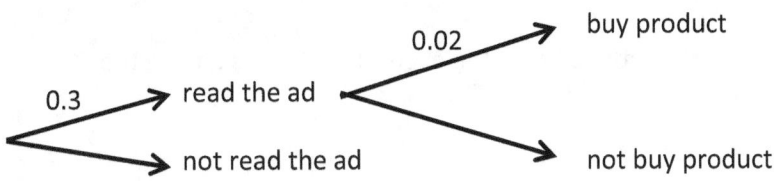

The required probability is (0.3)(0.02) = 0.006.

Problem 8: There are 5 coins in purse 1, 4 of which are S (silver) and 1 G (gold). There are 5 coins in purse 2, all of which are S (silver). Suppose that 4 coins are drawn at random from purse 1 and put into purse 2. Then 4 coins are drawn at random from purse 2 and put into purse 1. A person may now pick whichever purse he pleases. Which purse should he pick?

Purse 1 has: $S\,S\,S\,S\,G$	Purse 2 has: $S\,S\,S\,S\,S$

Solution 8: At the end, each purse contains the same number of coins, so he ought to pick the purse that has the greater probability to contain G (the gold coin). Now G can only be in purse 2 provided the following path is taken:

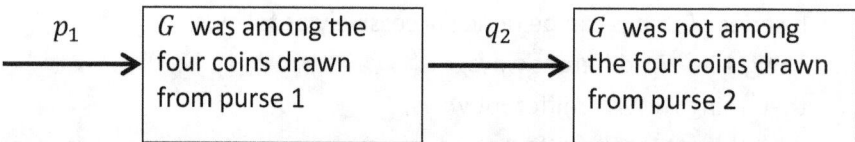

Instead of directly computing the probability of the event that G was among the 4 coins drawn from purse 1, let us first compute the probability of the complement event. The complement event is made up of one path, namely

$$4/5 \longrightarrow \text{not } G \longrightarrow 3/4 \longrightarrow \text{not } G \longrightarrow 2/3 \longrightarrow \text{not } G \longrightarrow 1/2 \longrightarrow \text{not } G$$

which has probability

$$q_1 = \frac{4}{5}\,\frac{3}{4}\,\frac{2}{3}\,\frac{1}{2} = \frac{1}{5}$$

Therefore the probability of the event that G was among the 4 coins drawn from purse 1 is

$$p_1 = 1 - q_1 = 1 - \frac{1}{5} = \frac{4}{5}$$

Given that G was among the 4 coins put into purse (so purse 2 now contains 8 S and 1 G), the event that G was not among the 4 coins drawn from purse 2 is made up of one path, namely

8/5 → not G → 7/8 → not G → 6/7 → not G → 5/6 → not G

which has the probability

$$q_2 = \frac{8}{9}\frac{7}{8}\frac{6}{7}\frac{5}{6} = \frac{5}{9}$$

Hence the probability that G is in purse 2 is

$$p_1 q_2 = \frac{4}{5}\frac{5}{9} = \frac{4}{9}$$

so the person should pick purse 1.

Tree diagrams

The multiplication rule can be displayed graphically by means of a tree diagram.

Problem: A man has 2 suits: gray and brown and 3 neckties: red, blue and green. What choice of suit and necktie does he have?

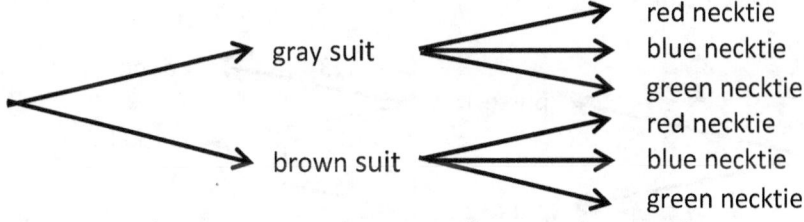

Solution: His choice is 2 times 3, or 6. To illustrate the use of a tree diagram, we let the left fork represent the choice of a suit. There are 2 suits so the left fork has 2 branches. Thus this fork is a 2-way fork. Each of the 2 branches on the left fork leads to a separate right fork. Each of the 2 right forks represents the choice of a necktie. There are three neckties so each right fork has 3 branches. Thus each right fork is a 3-way fork. A choice of both suit and necktie correspond to a choice of path from the left to right. As we see in the above tree diagram there are $2 \times 3 = 6$ such paths.

Problem: Purse 1 contains 1 dime and 2 nickels. Purse 2 contains 2 dimes and 1 nickel. You happen to take a purse and a coin from it. What is the probability that the coin is a dime?

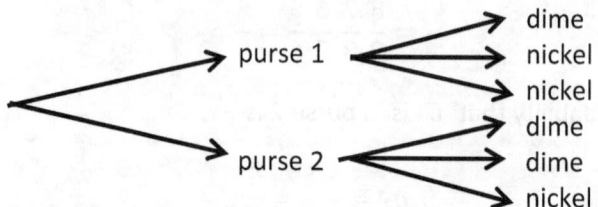

Solution: The first process is the taking of a purse; the second process is the taking of a coin from it. Both of these processes make up the overall process illustrated in the figure. We see there are 6 paths, each of which may be considered to represent 1 chance. Of these 6 chances, 3 result in a dime. Hence there are 3 chances in 6, or a probability equal to $3/6 = 1/2$, that the coin is a dime.

Alternative solution: Instead of constructing a tree diagram so that each path represents 1 chance, we can construct a tree diagram as shown in the figure below:

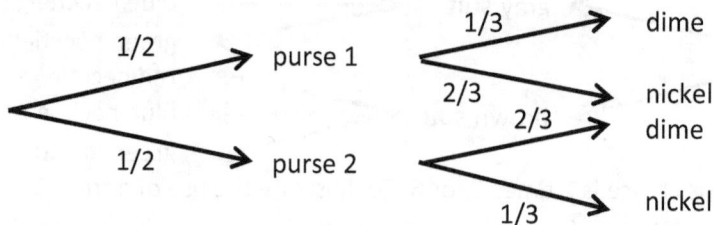

In this tree diagram, we have assigned a probability to each branch. For the first process, each purse has 1 chance in 2 to be taken, so we have assigned a probability of 1/2 to each branch. For the second process we must consider 2 situations, namely

(1) purse 1 was taken
(2) purse 2 was taken.

Given that purse 1 was taken, there is 1 chance in 3 for a dime, so the probability 1/3 is assigned to the *dime* branch leading out of purse 1.

Given that purse 1 was taken, there are 2 chances in 3 for a nickel. The probability 2/3 is assigned to the *nickel* branch leading out of purse 1.

Given that purse 2 was taken, there are 2 chances in 3 for a dime, so the probability 2/3 is assigned to the *dime* branch leading out of purse 2. Given that purse 2 was taken, there is 1 chance in 3 for a nickel, so the probability 1/3 is assigned to the *nickel* branch leading out of purse 2.

Let us consider the sequence of the two events

(1) purse 1 is taken
(2) a dime is taken from it.

This sequence of two events, which is the top path in the above figure, may be denoted by

$$\text{purse 1 then dime}$$

The probabilities on the branches of this path are:

(1/2) on the branch *purse 1*,
(1/3) on the branch *dime,* given *purse 1*.

Thus the probability of this path is

$$P(\text{purse 1 then dime}) = \left(\frac{1}{2}\right)\left(\frac{1}{3}\right) = \frac{1}{6}$$

In other words, the probability of the path corresponding to *purse 1* on the first process and *then dime* on the second process is equal to the product of the probabilities associated with each branch along the path. Likewise the probability of the path

$$\text{purse 2 then dime}$$

is

$$\left(\frac{1}{2}\right)\left(\frac{2}{3}\right) = \frac{2}{6}$$

The event of taking a dime is the union of the two events

purse 1 then dime,
purse 2 then dime,

which are incompatible. That is,

$$\text{dime} = (\text{purse 1 then dime}) \cup (\text{purse 2 then dime})$$

where

$$(\text{purse 1 then dime}) \cap (\text{purse 2 then dime}) = \emptyset$$

Hence by the addition rule for probabilities

$$P(dime) = P(purse\ 1\ then\ dime) + P(purse\ 2\ then\ dime)$$
$$= \frac{1}{6} + \frac{2}{6} = \frac{1}{2}$$

Exercises

1. A card is drawn from a well-shuffled pack of 52 cards. What is the probability of getting a king or a queen?

2. Two coins are tossed. What is the probability of *at least 1 T* (i.e. at least one tails)?

3. In how many ways can 3 letters be put into 3 addressed envelopes, one letter into each envelope?

4. A friend shows me 4 detective books and 8 western books, and lets me choose one of each. What choice do I have?

5. In how many ways can we select a consonant and a vowel out of the English alphabet?

6. There are 2 red balls (labeled $R1$, $R2$) and 3 black balls (labeled $B1$, $B2$, $B3$) in a bag. One ball is drawn. What is the probability that it is black?

7. The contents of the bag are the same as the foregoing problem, namely $R1$, $R2$, $B1$, $B2$, $B3$. For this problem, however, we suppose that one ball is drawn, its color is *unnoted,* and it is laid aside. Then another ball is drawn. What is the probability that the second ball drawn is B *given* that the first ball is *unknown? (Note:* We denote this probability by $P(B$ given? $)$ which is read *"P of B given unknown".)*

8. The contents of the bag are the same as the foregoing problem, namely $R1$, $R2$, $B1$, $B2$, $B3$. For this problem, however, we suppose that one ball is drawn, its color is *noted,* it is *red,* and it is laid aside. Then another ball is drawn. What is the probability that the second ball is *black, given* that the first ball drawn is *red? (Note:* We denote this probability by $P(B$ given $R)$ which is read *"P of B given R".)*

Chapter 9. Binomial distribution

"Strong reasons make strong actions." (*King John*).

Definition of binomial distribution
You would use binomial distributions in these situations:

(1.) When you have a limited number of independent trials, or tests, which can either succeed or fail
(2.) When success or failure of any one trial is independent of other trials

Binomial in probability begins with an action, or trial, having only two possible outcomes. The two and only two possible results of the action are called *success* and *failure*. Let the probability of a successful outcome of this action be the value p and the probability of failure be the value q. Because the sum of all the probabilities for an experiment is 1 and since there are only two outcomes, we have

$$p + q = 1 \text{ and } q = 1 - p$$

It does not matter which outcome is called a success and the other outcome is called a failure. The action might be the tossing a coin to obtain either heads or tails. For example, heads could be a success and tail a failure. The outcomes need not be equally likely. If the coin is fair, the probability of success is $p = 1/2$ and the probability of failure is $q = 1/2$. If the coin is unfair, the probability of success would not be the same as the probability of failure. For example, we could have $p = 0.3$, and $q = 1 - 0.3 = 0.7$.

We assume that each trial is independent. In other words, each trial stands alone. The result of one trial has no influence on any other trial. When a coin is flipped, the outcome is not influenced by the previous flip of the coin. A two-possible-outcome experiment, repeated a certain number of independent times, yields a binomial distribution. The distribution function has as a variable x, the number of successes. The other required parameters are the number n of independent trials, and the probability p of success on each trial. The probability of failure on each trial is $q = 1 - p$.

Let us summarize:

There are n identical trials. Each trial is independent; i.e., the result of one trial has no effect on another trial.

Each trial has only 2 outcomes; namely, success or failure. The probability of success on a trial is p . The probability of failure on a trial is q . Since there can only be success or failure, it follows that $q = 1 - p$ and $p = 1 - q$.

The number of successes in n trials is denoted by the variable x . The variable can only take on the numbers 0, 1, 2, 3, 4, and so on up to and including n .

Binomial trial

A Bernoulli trial (or binomial trial) is named after Jacob Bernoulli, a seventeenth-century Swiss mathematician. It is a random experiment with exactly two possible outcomes, "success" and "failure", for which the probability p of success is always the same. The labels "success" or "failure" for the two outcomes should not be construed literally. The term "success" means that one of two possible results occurred. The term "failure" means that the other result occurred. Specifically, a Bernoulli trial refers to whether a specific event occurs or not.

Because a Bernoulli trial has only two possible outcomes, it can be framed as a "yes or no" question. For example, the question might be, "Will it rain tomorrow?" The answer "yes" may be called a success and the answer "no" may be called a failure. Examples of Bernoulli trials include: (1) Flipping a coin, where heads denotes success and tails denotes failure. A fair coin has the probability of success 0.5 by definition. (2) Rolling a die, where a six is "success" and any one of the oter five numbers is a "failure". In this case there are six possible outcomes, and the event is a six; the complementary event "not a six" corresponds to the other five possible outcomes.

A sequence of independent Bernoulli trials with probability 1/2 of success on each trial is called a fair coin. One for which the probability is not 1/2 is called a biased or unfair coin. John Edmund Kerrich performed experiments in coin flipping and found that a coin made from a wooden disk about the size of a quarter-dollar coin and coated on one side with lead landed heads (wooden side up) 679 times out of 1000. Such a coin was unfair.

If someone has altered a coin to prefer one side over another (a biased coin), the coin can still be used for fair results by changing the game slightly. John von Neumann gave the following procedure:

1. Toss the coin twice.

2. If the results match (i.e. either HH or TT), start over, forgetting both results.

3. If the results differ (i.e. either HT or TH), use the first result (i.e. either H or T), forgetting the second i.e. either T or H).

The reason this process produces a fair result is that the probability of getting heads and then tails (i.e. HT) must be the same as the probability of getting tails and then heads (i.e. HT). By excluding the events of HH and TT tails by repeating the procedure, the coin flipper is left with the only two remaining outcomes HT and TH having equivalent probability.

Drawing a playing card

The expression "luck of the draw" refers to *pure chance*, as in the sentence, "It isn't anyone's fault; it's just the luck of the draw." *This expression alludes to the random drawing of a playing card.*

The primary deck of 52 playing cards has thirteen ranks of each of the four suits: red diamonds (♦), black spades (♠), red hearts (♥) and black clubs (♣).A face card is a king, queen, or jack of a suit. Half of the cards are black and the other half are red. Each suit of 13 cards has

One ace card, representing the number 1;
Nine cards numbered 2 through 10;
Three face cards (king, queen, and jack).

The basic trial is the drawing of one card from a shuffled deck. The card is observed and then returned to the deck. The deck is reshuffled making it ready for the next trial. This process is repeated tor each trial. The probabilities of black card and red card on a single trial are the same; namely,

$$p = P(B) = \frac{26}{52} = 0.5$$

$$q = P(R) = \frac{26}{52} = 0.5$$

A player performs two trials. What is the chance of one black card and one red card in either order? The two trials may consist of two red cards, or red card and black card, or two black cards. These three possibilities, however, are not equally likely. What we really have is four equally-likely possibilities, which are:

red card-red card,
red card-black card,
black card-red card
black card-black card.

Because two of these four possibilities consist of a red card and a black card, there are two cases out of four that the player has a mixed pair in two trials, which is a probability of 0.5.

The Unicode for playing cards provide for three jokers: red, black, and white. Let us now suppose that a black joker has been added to the deck of 52 cards. Thus there are 53 cards in the deck, of which 27 are black and 26 are red. Let B stands for black card and R stands for red card. The probabilities of black card and red card are now

$$p = P(B) = \frac{27}{53} = 0.509434 \ldots \approx 0.51$$

$$q = P(R) = \frac{26}{53} = 0.490566 \ldots \approx 0.49$$

For simplicity we have rounded these two probabilities to 0.51 and 0.49. That one trial has no effect on any other trial is the notion of independence which is basic to the development we give here. The first trial can result in either black card or red card, with the probabilities 0.51 or 0.49 respectively. Thus, regardless of the outcome of the first trial, the second trial must result in either black card or red card with the same probabilities (i.e., 0.51 and 0.49) respectively, Combining the possible outcomes of the second trial to those of the first trial we obtain:

RR	(0.49)(0.49)
RB	(0.49)(0.51)
BR	(0.51)(0.49)
BB	(0.51)(0.51)

Here RR stands for red card in the first trial and red card in the second trial and (0.49)(0.49) is the probability of this event. Here we have used the rule of the multiplication of probabilities. Likewise RB stands for red card-black card and (0.49)(0.51) is the probability of this event. The remaining two entries are similarly interpreted. The above table shows that the combination of two trials leads to four possible outcomes. However, we can combine the two outcomes RB and BR because they have the common property that each contains one red card and one black card. If one does not consider the order in which R and B are obtained, the outcomes RB and BR are identical, Moreover we see that each of the events RB and BR has the same probability, namely (0.51)(0.49). From this point of view only three separate outcomes are possible: (1) neither trial results in a black card, (2) only one trial results in a black card, or (3) both trials result in black cards. We can summarize this new way of looking at the results of two trials as follows:

$$P(0B) = (0.49)^2$$
$$P(1B) = 2(0.51)(0.49)$$
$$P(2B) = (0.51)^2$$

Here $0B$ stands for zero black cards and two red cards and $(0.49)^2$ is the probability of this event. Likewise $1B$ stands for one black card and one red card and $2(0.51)(0.49)$ is the probability of this event. This probability has resulted from use of the addition rule, for we added the probability $(0.51)(0.49)$ of RB and the probability of $(0.51)(0.49)$ of BR to obtain $2(0.51)(0.49)$. Finally, $2B$ stands for two black cards and no red cards and $(0.51)^2$ is the probability of this event.

Let us go on now and consider a third trial. Considering all possible outcomes and their probabilities we obtain the table:

RRR	$(0.49)^3$
RRB	$(0.51)(0.49)^2$
RBR	$(0.51)(0.49)^2$
RBB	$(0.51)^2(0.49)$
BRR	$(0.51)(0.49)^2$
BRB	$(0.51)^2(0.49)$
BBR	$(0.51)^3$
BBB	$(0.51)^3$

The above columns show eight cases. However we may bring together all the cases leading to the same red-black make-up to obtain the table:

$0B$	$(0.49)^3$
$1B$	$3(0.51)(0.49)^2$
$2B$	$3(0.51)^2(0.49)$
$3B$	$(0.51)^3$

For example the event $2B$ is made up of the three cases RBB, BRB, and BBR each of which has a probability $(0.51)^2(0.49)$. Thus the probability of the event $2B$ is $3(0.51)^2(0.49)$.

Without going into further details we can now see the simple mechanism by which one obtains the probabilities for the various arrangements in a series of any number of trials. For example, let us consider we have four trials and we wish to find the probability of the event of two black cards and two red cards. This event is made up of 6 cases, namely

$$RRBB, RBRB, RBBR, BRRB, BRBR, BBRR$$

The probability of each of these cases is $(0.51)^2(0.49)^2$.. Since these six cases make up the event, the probability of the event $2B$ is $6(0.51)^2(0.49)^2$. Here we found the number 6 by actually listing the cases which had two black cards and two red cards in any order. However there are easier ways to find the number of cases for any event.

Pascal's triangle

One such way is by use of *Pascal's Triangle*. In the first row one writes the number 1, and in the second row the numbers 1 and 1 so that the three numbers form a triangle. Then one constructs additional rows of this triangle by adding together two adjacent numbers in a row and placing the sum in the next row down at a location between the two adjacent numbers. In the application of this rule it is assumed that rows are extended by zeroes to the left and to the right. Thus Pascal's Triangle for the first six rows is

row 0						1					
row 1					1		1				
row 2				1		2		1			
row 3			1		3		3		1		
row 4		1		4		6		4		1	
row 5	1		5		10		10		5		1

The top row is labeled 0, the next is row 1, the next row 2, etc. We see that row three, namely, 1, 3, 3, 1, gives for three trials the number of cases for the events $0B$, $1B$, $2B$, and $3B$ respectively. Likewise, row number four, namely, 1, 4, 6, 4, 1, gives for four trials the number of cases for the events $0B$, $1B$, $2B$, $3B$, and $4B$ respectively.

This table can be easily continued. Row number 6 is obtained by

$$0+1, 1+5, 5+10, 10+10, 10+5, 5+1, 1+0$$

which is

$$1, 6, 15, 20, 15, 6, 1$$

Because of these calculations, we see that the sum of the numbers in row 6 is 64. The sum of numbers in row 5 is 32. We note that 64 is twice 32. In fact, we have

$$\text{Sum of Numbers in Row } 0 = 2^0 = 1$$
$$\text{Sum of Numbers in Row } 1 = 2^1 = 2$$
$$\text{Sum of Numbers in Row } 2 = 2^2 = 4$$
$$\text{Sum of Numbers in Row } 3 = 2^3 = 8$$
$$\text{Sum of Numbers in Row } 4 = 2^4 = 16$$

$$\text{Sum of Numbers in Row } n = 2^n$$

The various rows of the table are symmetrical. In rows corresponding to an even number of trials, say row 4, the numbers increase toward the middle, and the greatest number occurs at the middle corresponding to an equal number of black cards as red cards. In rows corresponding to an odd number of trials, the numbers also increase toward the middle, but now there are two equal numbers in the middle corresponding to the two cases where there is one more black card than red card or one more red card than black card.

Each number in Pascal's Triangle can be represented by the symbol $\binom{n}{x}$. This symbol represents the number of combinations of n things taken

x at a time. This symbol is made up of two numbers one over the other and both enclosed by a set of parentheses. The top number is the number of trials and the bottom number is the number of black cards. With this symbol Pascal's Triangle becomes

$$\binom{0}{0}$$

$$\binom{1}{0} \qquad \binom{1}{1}$$

$$\binom{2}{0} \qquad \binom{2}{1} \qquad \binom{2}{2}$$

$$\binom{3}{0} \qquad \binom{3}{1} \qquad \binom{3}{2} \qquad \binom{3}{3}$$

$$\binom{4}{0} \qquad \binom{4}{1} \qquad \binom{4}{2} \qquad \binom{4}{3} \qquad \binom{4}{4}$$

$$\binom{5}{0} \qquad \binom{5}{1} \qquad \binom{5}{2} \qquad \binom{5}{3} \qquad \binom{5}{4} \qquad \binom{5}{5}$$

$$\binom{6}{0} \quad \binom{6}{1} \quad \binom{6}{2} \quad \binom{6}{3} \quad \binom{6}{4} \quad \binom{6}{5} \quad \binom{6}{6}$$

which is

row 0:			1			
row 1:		1		1		
row 2:	1		2		1	
row 3:	1	3		3		1
row 4:	1	4		6	4	1

row 0:   1
row 1:   1   1
row 2:   1   2   1
row 3:   1   3   3   1
row 4:   1   4   6   4   1
row 5:   1   5   10   10   5   1
row 6: 1   6   15   20   15   6   1

Permutations and combinations

In order to understand combinations we must first discuss permutations. Each different arrangement either of all or part of a number of things is called a *permutation*. Thus there are two permutations of the two letters A and B, namely AB and BA. There are six permutations of the three letters A, B, and C, namely ABC, ACB, BAC, BCA, CAB, CBA. There are also six permutations of the three letters A, B, and C taken two at a time, namely AB, AC, BA, BC, CA, CB.

Let us now find an expression for the number of permutations of n things taken x at a time, symbolized by P_x^n. The first place can be filled in any one of n ways. This leaves $n - 1$ ways of filling the second place by any one of the $n - 1$ remaining things. The third place can be filled in $n - 2$ ways, the fourth in $n - 3$ ways, and so on until finally the xth place in $n - x + 1$ ways. Hence the number of ways of filling the x places is

$$P_x^n = n(n - 1)(n - 2)(n - 3) \cdots (n - x + 1)$$

We observe that this expression can also be written as

$$P_x^n = \frac{n!}{(n - x)!} = \frac{n(n - 1)(n - 2)(n - 3) \cdots (3)(2)(1)}{(n - x)(n - x - 1)(n - x - 2) \cdots (3)(2)(1)}$$

If $x = n$ then the expression for the number of permutations is

$$P_n^n = n(n - 1)(n - 2)(n - 3) \cdots (3)(2)(1) = n!$$

That is, the number of permutations of n things taken n at a time is $n!$ (which is called n factorial).

Some slight knowledge of permutations seems to have existed among the ancient Chinese. Although the Greeks gave the subject little attention, the Latin mathematician Boethius in about 510 AD gave the rule for the number of permutations of n things taken two at a time; that is,

$$P_2^n = n(n - 1)$$

Finally, about 1150 AD, Bhaskara gave the general rule for both permutations and combinations.

A combination is a group of things that is independent of the order of the various distinct elements in the group. For example, the musical effect of striking three notes at the same time on a piano, say C, E, G, is the same whether we think of the notes as in one order or another.

This is a combination of notes. If, however, we strike the same notes in succession, the musical effect is different when the notes are struck in different orders. This is a permutation of notes.

Let us now find an expression for $\binom{n}{x}$, the number of combinations of n things taken x at a time. Since any change of order of the elements in a combination does not alter the combinations, it follows that in any one combination of x things there are $x!$ permutations. Hence there are $x!$ times as many permutations in a given group as there are combinations. That is,

$$P_x^n = x! \binom{n}{x}$$

or

$$\binom{n}{x} = \frac{P_x^n}{x!} = \frac{n(n-1)(n-2)(n-3)\cdots(n-x+1)}{x(x-1)(x-2)(x-3)\cdots(3)(2)(1)}$$

In the application of this formula we observe that there are the same number of factors in the numerator as there are in the denominator. An alternative expression is

$$\binom{n}{x} = \frac{n!}{x!\,(n-x)!}$$

As an example, the number of cases of 4 trials resulting in 3 black cards is the number of combinations of 4 things taken 3 at a time; namely,

$$\binom{4}{3} = \frac{4!}{3!\,1!} = \frac{4\cdot3\cdot2\cdot1}{3\cdot2\cdot1\cdot1} = 4$$

which is the same result that we can obtain from Pascal's Triangle.

The binomial distribution

We can summarize our results up to this point by saying that in n trials the number x of black cards follows a binomial distribution. As we have seen the probability of each case with x black cards and $n-x$ red cards is

$$(0.51)^x (0.49)^{n-x}$$

and there are $\binom{n}{x}$ such cases. Hence in n trials the probability of x black cards is

$$P(x) = \binom{n}{x}(0.51)^x (0.49)^{n-x}$$

This equation represents the *binomial distribution*, where the number x of black cards can take any value $x = 0, 1, 2, \cdots, n$. We have already written down the distributions when $n = 1, 2,$ and 3, which were

$n = 1$	$P(0) = 0.4$
	$P(1) = 0.51$
	Total $= 1.00$
$n = 2$	$P(0) = (0.49)^2 = 0.24$
	$P(1) = 2(0.51)(0.49) = 0.50$
	$P(2) = (0.51)^2 = 0.26$
	Total $= 1.00$
$n = 3$	$P(0) = (0.49)^3 = 0.12$
	$3(0.51)(0.49)^2 = 0.37$
	$3(0.51)^2 2(0.49) = 0.38$
	$(0.51)^3 = 0.13$
	Total $= 1.00$

Let us now change our terminology from that of drawing cards to the general terminology for the binomial distribution.

Instead of a drawing we speak of a trial of an experiment. Each trial has two possible outcomes; namely, success or failure. The probability of a success on a given trial is denoted by p and the probability of a failure on the given trial is denoted by q. Since a trial must result in either a success or a failure we have $p + q = 1$. In our case success corresponds to the trial of a black card B and the probability of a success is $p = 0.51$. Failure (and the use of the word is only as an expression) corresponds to the trial of a red card G and the probability of failure is

0.49. The binomial distribution gives for n trials the probability of x successes. This probability is

$$P(x) = \binom{n}{x} p^x q^{n-x}$$

where x can vary over the range $0, 1, 2, \cdots, n$. It is customary to say that the number of successes in n trials is a random variable x having the binomial distribution. This distribution applies whenever the probability of a success remains constant from trial to trial and the trials are independent.

Mean of the binomial distribution

Let us now apply the concept of expectation to the binomial distribution. First consider the distribution of the number x of black cards in the case of one trial:

$$x = 0 \qquad P(x = 0) = 0.49$$
$$x = 1 \qquad P(x = 1) = 0.51$$

We multiply the number of black cards by the corresponding probability to find the expectation of each event. By adding all these products we obtain the expected number of black cards:

$$0 \times P(0) + 1 \times P(1) = 0.51$$

Thus in one trial the expected number of black cards is 0.51. We call this expected number the mean of the binomial distribution for $n = 1$. Note that this mean is equal to $np = (1)(0.51)$.

Next consider the binomial distribution for $n = 2$ trials:

number of black cards	probability	product
$x = 0$	$P(x = 0) = 0.24$	0.00
$x = 1$	$P(x = 1) = 0.50$	0.50
$x = 2$	$P(x = 2) = 0.26$	0.52

We take each product of the number of black cards times its probability. We sum these products to find the expectation:

$$0.50 + 0.52 = 1.02$$

Thus the mean number of black cards in 2 trials is 1.02. Note that this mean is equal to

$$np = 2(0.51) = 1.02$$

Now consider the binomial distribution for $n = 3$ trials:

Number of black cards	probability	product
0	0.12	0.00
1	0.37	0.37
2	0.38	0.76
3	0.13	0.39
		Total = 1.52

The expected number of black cards in 3 trials is the sum of the products, namely 1.52. Because we rounded off our probabilities to two decimal places, there is round-off error in our expected value. If we had carried more decimal places, the expected value would have been 1.53, which is *np* = 3(0.51).

We see that the *mean of a binomial distribution* is equal to *np*. In other words the mean of a binomial distribution is equal to the product of the number of trials and the probability of success on an individual trial. Another term for the mean is the expected value of the number of successes. Hence we may write that for a binomial distribution

$$Ex = np$$

which is an abbreviation for "the expected value of x is equal to np." Often we denote the mean of a probability distribution by the Greek letter μ. Hence we may also write that for a binomial distribution

$$\mu = np$$

We have not proved this formula that the mean of a binomial distribution is given by np, but the formula is intuitively reasonable. For example, in 100 trials we would expect $np = 100(0.51) = 51$ black cards.

In this section we have seen how to find the mean or expected number of black cards in n trials. This mean is a measure of the central location, and is the counterpart of the mean or average of a set of data. In fact the formula for the *empirical mean* is

$$\bar{x} = \sum xf$$

where x represents the variable and f represents its relative frequency. The formula for the theoretical mean IL is

$$\mu = \sum xP$$

where x represents the random variable and P represents its probability. Even as the relative frequency f is an estimate of the probability P, so is the empirical mean $\bar{x}$ an estimate of the theoretical mean μ. In the case of the binomial distribution we have exact expressions for the probability values P and hence we can theoretically compute the value of μ which turns out to be np.

Variance of the binomial distribution
Previously we have seen how to describe the variability of data expressed in terms of an empirical frequency distribution. In particular the *empirical variance* is given by the formula

$$s^2 = \frac{1}{n-1} \sum (x - \bar{x})^2 \, f$$

where f is now the absolute frequency (i.e., the numbers of data points in each class interval). In case the theoretical mean μ is known, we should instead use the formula

$$s^2 = \frac{1}{n} \sum (x - \mu)^2 f$$

That is, when the empirical mean $\bar{x}$ is used, the divisor in s^2 is $n - 1$, whereas if the theoretical mean μ is used the divisor is n. Since relative frequency is absolute frequency divided by n, this last formula in terms of relative frequency is simply

$$s^2 = \sum (x - \mu)^2 f$$

The formula for the theoretical variance is

$$\sigma^2 = \sum (x - \mu)^2 P$$

where P is the counterpart of f, and σ^2 is the counterpart of s^2. The quantity σ^2 is called the *(theoretical) variance*. It is represented as the square of the Greek lower case letter sigma σ. The theoretical variance may also be described as the expected value of $(x - \mu)^2$. That is,

$$\sigma^2 = E(x - \mu)^2$$

Let us now compute the variance of the binomial distribution for one trial. We have $n = 1$ and $\mu = 0.51$.

Number of black cards	Probability	Squared Deviations	Products
0	0.49	$(0 - 0.51)^2$	0.127449
1	0.51	$(1 - 0.51)^2$	0.122451
			Total = 0.249900

The total of the products of squared deviations × probability is

$$\sigma^2 = 0.51 \times 0.51 \times 0.49 + 0.49 \times 0.49 \times 0.51$$
$$= (0.51 + 0.49) \times 0.51 \times 0.49 = 0.2499$$

We see that this value is $npq = (1)(0.51)(0.49) = 0.2499$.

Let us next compute the variance of the binomial distribution of two trials. We have (where $n = 2$, $\mu = 1.02$)

x	P	$(x - \mu)^2$	$(x - \mu)^2 P$
0	0.24	$(0 - 1.02)^2$	0.25
1	0.50	$(1 - 1.02)^2$	0.00
2	0.26	$(2 - 1.02)^2$	0.25
			Total =0.50

The total is the variance $\sigma^2 = 0.50$ which again we see is equal to $npq = 2(0.51)(0.49)$.

Finally, the binomial distribution for $n = 3$ trials is (where $\mu = 1.53$):

x	P	$(x - \mu)^2$	$(x - \mu)^2 P$
0	0.12	$(0 - 1.53)^2$	0.281
1	0.37	$(1 - 1.53)^2$	0.104
2	0.38	$(2 - 1.53)^2$	0.084
3	0.13	$(3 - 1.53)^2$	0.281
			Total = 0.750

The total is the variance as $\sigma^2 = 0.750$, which again we see is $npq = 3(0.51)(0.49)$. In general we can say that the *mean* μ of a *binomial distribution is* np and the *variance* σ^2 *is* npq. As a result the *standard deviation* of a binomial distribution is the positive square root of the variance: $\sigma = \sqrt{npq}$.

The standard deviation of the above three distributions (where $pq = 0.51 \times 0.19 = 0.2499 \approx 0.25$) are:

For $n = 1$	$\sigma = \sqrt{0.25} = 0.50$
For $n = 2$	$\sigma = \sqrt{0.50} = 0.71$
For $n = 3$	$\sigma = \sqrt{0.75} = 0.87$

Independence between individual trials
During the last century in some parts of Europe a man whose wife was expecting a baby would be extremely anxious for a black card. He would go to the registrar of trials in the town to see how many black cards and how many red cards were born in the last few days. If the number of red cards was comparatively large, he would be happy and would feel that his probability of having a son instead of a daughter was exceedingly good.

Because of the great interest in this problem, statistical evidence was collected. Statistical data from four towns in Bavaria showed that out of 200,000 trials there was only one run of 17 consecutive trials of babies of the same sex, and not a single such run of greater length.

Many people interpret this statistical data as follows. Once 17 red cards are born in succession in a town, the probability that the next baby born is a black card is so great that the event is practically a certainty. Thus they reason: After a run of 17 consecutive red card trials, the next trial is for all intents and purposes a sure thing, namely the certain trial of a boy, instead of a random event.

But what about the man who was the father of the 17th red card in this run of 17 consecutive red cards? Before his daughter was born, he also went to the town registrar and saw that there was a run of 16 consecutive female trials. Was he not happy also, and did he not consider that his probability of a male trial was exceedingly good? And yet fate decreed otherwise, and his wife had a red card, despite the fact that the 16 preceding trials were also red cards.

As more and more statistical evidence is collected and assimilated, the following characteristic stands out. Let us denote the trial of a black card by B and trial of a red card by G. Then the successive trials would be represented by a series, such as

$$\dots BBGGGGBGBGGBGBBBG \dots$$

The statistical evidence shows that the distribution of B's and G's in the partial series formed by all trials following 17 consecutive G's is the same as the distribution in the whole series. Consequently on the basis of this evidence, the probability of a black card after a run of 17 consecutive red cards is the same as the probability of a black card regardless of what trials came before. In other words, the probability of a black card is not altered by the event of the 17 preceding trials all being red cards.

Let G^{17} designate the event of 17 consecutive red cards. Let $B|G^{17}$ designate the event of a black card given that the 17 preceding trials were all red cards. (Note: The vertical bar I is read *given that*, or simply *given*.)   Let B designate the event of a black card. Now let X designate any arbitrary event. Then we let $P(X)$ designate the probability that the event X happens. Thus

$$P(G^{17})$$

is the *probability* that the event "17 consecutive red cards" happens, and

$$P(B|G^{17})$$

is the *probability* that the event "a black card given that the 17 preceding trials were red cards" happens. And finally

$$P(B)$$

is the *probability* that the event "a black card" happens. Then the empirical experience of mankind, based on statistical evidence collected over the centuries, indicates the following relationship holds

$$P(B) = P(B|G^{17})$$

That is, the probability of a black card is the same as the probability of a black card given that the 17 consecutive prior trials were red cards. Thus we may say that the event "a black card" is independent of the event of 17 consecutive prior red card trials.

Exercises

1. Seven horses are in a race. In how many ways can three from among them finish first, second, and third? *Answer.* 210

2. How many five-card hands can be dealt from a 52-card deck? *Answer.* 2,598,960

3. How many three-letter words can be made from ten different letters (a) if repeats are allowed? (b) if repeats are not allowed? (c) if a letter may be repeated only once? *Answer.* (a) 1000 (b) 720 (c) 990

4. (a) In how many ways can a president, a secretary, and a treasurer be selected from a club of 20 people? (b) In how many ways can a committee of three be selected from the club? *Answer.* (a) 6840 (b) 1140

5. In how many ways can the 26 letters of the alphabet be lined up so that A and B are adjacent? *Answer.* 2×25!

6. (a) How many 13-card bridge hands can be dealt from a deck of 52 cards? (b) In how many ways can 13-card hands be dealt to North, South, East, and West?

Answer (a) $\binom{52}{13}$ (b) $\binom{52}{13}\binom{39}{13}\binom{26}{13}\binom{13}{13}$

7. How many five-card poker hands consist of (a) two pairs? (b) a full house? (c) a straight flush? (d) four of a kind?

Answer (a) $\binom{13}{2}\binom{4}{2}\binom{4}{2} 44 = 123,552$

Answer (b) $13\binom{4}{3} 12\binom{4}{2} 44 = 3774$

Answer (c) $4 \times 10 = 40$

Answer (d) $13 \times 48 = 624$

6. A lock has 40 positions. A "combination" for the lock consists of four settings, and no setting can coincide with the preceding one. How many such "combinations" are there? *Answer.* (40)(39)(39)(39) = 2,372,760

9. Six people sit in six chairs in a circle. If everyone moves the same number of places to the left, the seating is considered the same as before. How many different seatings are there? *Answer.* 6!/6 = 120.

10. An entering college student must take one of five science courses, one of six history courses, one of five English courses, and one of three mathematics courses. How many programs are available to him? *Answer.* (5)(6)(5)(3) = 450

11. An advertising agency claims that, of the college students who smoke, 25 percent smoke brand A. If we take a random sample of four students who smoke, and if the claim is true, what is the probability that at least one of them will be found to smoke brand *A? Answer.* 175/256.

12. In Boston, rain falls on the average one day out of every three days during which the sky is overcast. What is the distribution of the number of days with rainfall among the next three overcast days, assuming complete independence? Find its mean and variance. *Answer*

$$P(x) = \binom{3}{x}\left(\frac{1}{3}\right)^x\left(\frac{2}{3}\right)^{3-x}, \qquad x = 0,1,2,3$$

mean=1, variance=2/3

13. In a large orchard, 10 percent of the apples have worms. If four apples are picked at random, what is the probability that (a) exactly one will be wormy? (b) none will be wormy? (c) at least one will be wormy?

Answer. (a) 0.2916 (b) 0.6561 (c) 0.3439

Chapter 10. Population and sample

"Things won are done; joy's soul lies in the doing." (*Troilus and Cressida*)

Population

Any set of individuals, objects, or ideas that have some common observable characteristic makes up a *population* or *universe*. For example, the set of all people living in America would be a population of individuals; the set of all electric light bulbs in a warehouse would be a population of objects; the set of all poker hands (i.e. a set of 5 cards dealt from a shuffled deck) is a population of ideas. The notion common to all populations is that of aggregation; the term population is employed to denote any collection of a specific type under consideration.

A population can be finite or infinite. The population of people in America is finite. The population of electric light bulbs in a warehouse is finite. Moreover the population of people in America is existent; that is, the population exists in the sense that we can go out and count each member of the entire population. Likewise the population of electric light bulbs is existent; only lack of time, money or opportunity would prevent us from examining the whole population. On the other hand, the population of all poker hands is not finite. Each time we gather up the cards, reshuffle the deck, and deal a new hand we generate another member of the population. Now it is clear that no matter how many times we deal we can never obtain the complete population, for we can always add new members by another deal. As a result this population is infinite. Furthermore, the population does not exist in the sense that we can go out and look at any member. Instead the members have a kind of hypothetical existence conferred on them by the notion of the process of dealing a hand.

With the above comments in mind, we may say that generally a population is an abstract set, which in certain specific cases actually reduces to a concrete set. Let us give some further examples: the population of a shipload of grain and the population of points on a rifle target. The shipload of grain is finite but so large that we usually would treat it as infinite; the points on a target are indeed infinite. Most of our procedures will be based on the assumption of an infinite population.

One definition of the science of *statistics* is that it is the branch of scientific method which deals with the properties of populations. More specifically statistics deal with the numerical properties of populations, so with each member of the population we associate a number called the measurement. Depending on the characteristic at hand, the numbers can be continuous or discrete. For example, if we consider the height of people in America the variable "height" is a continuous variable; it can take on all possible values from say one-half to eight feet. In dealing with continuous variables we never look at one particular value, say five feet eight inches, but at a small interval around a particular value, say five feet eight inches give or take one-half inch. In other words there is always some region of tolerance associated with any measurement of a continuous variable. As another example, if we consider the number of children that each women has in America, the variable "number of children" is a discrete variable; it can take on only the discrete values 0, 1, 2, . . . , 20 or whatever the upper limit might be.

Sample

Any subset of a population is a *sample* from that population. The *size of a sample,* usually represented by the letter n, is the number of members in the sample. A field of corn has an area of 1000 acres. A sample of 20 plots of one acre each is chosen, and the yields of these plots are measured. The population consists of the 1000 individual acres as members. The sample is the 20 members chosen, while the measurement is the yield of the plot.

An investigation is made to test the effects of a certain type of vaccine for flu. A group of subjects infected with flu are treated with the vaccine, and the number recovering from the flu within a specified time interval is observed. The sample consists of the group of subjects actually used, and the measurement would be the label either "recovered" or "did not recover." The population here could be considered to be all people with the flu.

A study is made as to the age of people in the United States. A sample of 1000 people is chosen, and their ages recorded. Here the population consists of all people in the United States. The sample consists of the 1000 people chosen. The measurement taken is age.

Random sampling

A critical factor in finding out what use can be made of a sample is the method used in choosing the sample. If some members of the population have a greater chance to be chosen than others, then we say the sample is biased. For example, suppose we are sampling people to determine the average age of a population. If each time we came upon an older looking person we neglected him and instead picked a younger looking person, then our sample would be biased toward younger people. Such subjective methods of picking members from a population for a sample often result from subconscious or sometimes conscious preferences of the person making the selections. In order to prevent such bias, an objective method of picking a sample ought to be employed.

When every member of the population has an *equal* and *independent* chance of being chosen for a sample, the sample is called a *random sample.* In other words, a random sample is one in which (1) all members of the population have an equal opportunity to be drawn into the sample, and (2) each member is selected independently of whether any other member is drawn into the sample. Taken together, these conditions mean that any group of members is as likely to be chosen as any other group of the same size.

Technically speaking, every member chosen should be measured and returned to the population before another selection is made. As a result it is possible that some member can be chosen twice in the same sample. For example, to choose a random sample of 5 cards out of a population of 52 cards, you ought to choose one card, record its value, return the card to the deck, shuffle the deck, then draw the second card, record its value, return it to the deck, shuffle, and so on until 5 cards have been recorded. If the population is large compared to the sample size, then only a small error will result from the procedure of not returning each sampled member back to the population. In practice it is usual not to return each sampled member, and in fact sometimes it is impossible to do so, as in a case where a member is changed or destroyed in the sampling process.

Theoretical frequency curve (for a finite population)

Note that we said in our definition of random sample that every member of the population has an equal chance of being in the sample. We did not say, and it is wrong to say, that every measurement in the population has an equal chance of being in the sample. For example, in the U.S. population there are many more people whose height measures 5'9" than 6'9". In a random sample each member has an equal opportunity, and as a result the measurement 5'9" will have a greater chance of being drawn than 6'9". In fact the curve that represents the probabilities of the measurements for a single draw (i.e. a sample of size one) is the *theoretical frequency curve* for heights in the population.

The theoretical frequency curve is an abstraction from the empirical frequency curve of a set of data. If the set of data makes up the measurements of the entire population (as it can only in the case of certain finite population) then the theoretical and empirical curves are the same. For example, the U.S. Census each ten years records the age of each person in the United States. The empirical frequency curve of ages for this data, because it exhausts the entire U.S. population, is the theoretical distribution as well. The median of this distribution is that age for which half the people are older and the other half younger. Now we are going to make an important jump ahead in our statistical logic. We are going to say that if we draw a person at random (that is, if we take a random sample of size one) then there is a fifty percent chance that this person will be older than the median age and a fifty percent chance this person will be younger than the median age. In other words, our statistical logic is this: Fifty percent of the people in the population are older than the median age; therefore, a person drawn at random has a fifty percent chance of being older than the median age. The above reasoning is general; the median is associated with 50 percent, but one can use any other benchmark with its associated percent. For example, the concept can be extended from one value (the median) which divides the total frequency into two equal parts, to an arbitrary value which divides the total frequency into two unequal parts. The percentage that lies above this value becomes the chance that a random draw exceeds this value. Likewise the percentage that lies below this value becomes the chance that a random draw falls short of this value.

Theoretical frequency curve (for an infinite population)
Let us consider a simple example. Suppose that a coin is tossed a number of times and frequencies of heads and tails are recorded. The relative frequencies of each can be shown in a histogram. In the figure we show three histograms, one for 1 throws, one for 100 throws, and one for 1000 throws. Note that on the abscissa we represent tails by 0 and heads by 1.

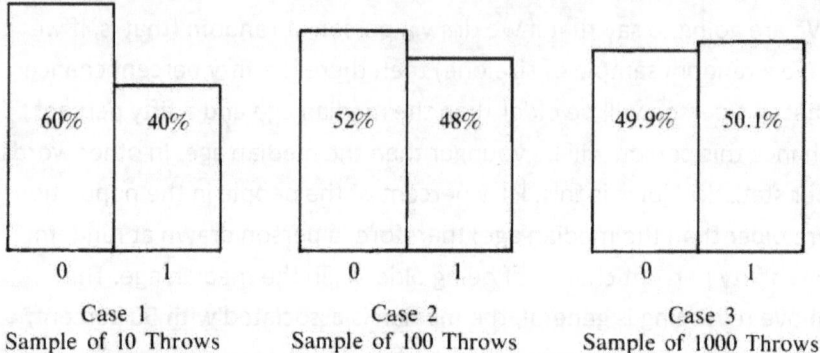

Case 1	Case 2	Case 3
Sample of 10 Throws	Sample of 100 Throws	Sample of 1000 Throws

As the number of throws is increased so in the regularity of the histogram, and the frequency distribution of the sample is said to tend to the frequency distribution of the population. In this case, for an honest coin, the frequency distribution of the population is 50%–50%. Note here that the *a priori* (i.e. beforehand) knowledge of the population distribution allows us to predict with reasonable accuracy the behavior of the sample.

As a further demonstration, let us take the total score when 10 coins are thrown at the same time. This score cannot be less than zero or more than 10. Again, the frequency distribution of the population can be determined by using the methods of probability theory. The figure shows how the histograms become more regular as the throws increase, and tend to the frequency distribution of the population:

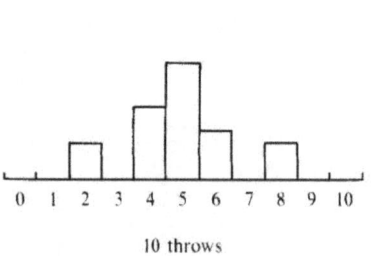

10 throws

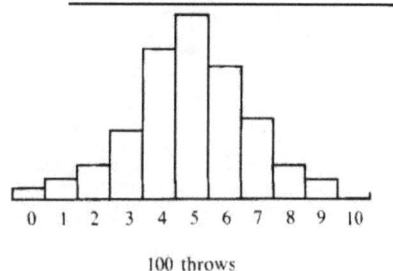

100 throws

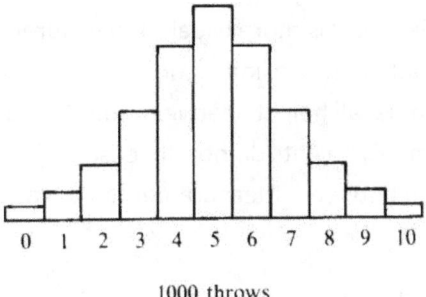

1000 throws

The essential feature of a histogram is that the proportion of occasions on which the score is between any two values is represented by the area of the blocks of those two values and all the blocks in between.

A more practical example is shown in the figure which gives histograms for heights of American men:

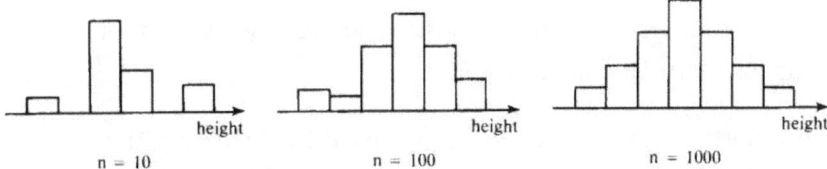

As the size of the sample is increased the irregularities of the histogram disappear and the form of the frequency curve of the population becomes more apparent. Of course the distribution is not determined *a priori* (beforehand) as with the distribution of scores of coin tossing. However this distribution of heights and the distribution from coin tossing bear marked similarities. This similarity is not purely coincidental, but it is a consequence of the *central limit theorem* which determine the shape of many frequency distributions experienced in practice; namely, the *normal curve.* More on the central limit theorem and the normal curve will be given in the next chapter.

In the examples given on the tossing of coins, the scores must be whole numbers. Such cases where the measurement must be a whole number do occur, as for example, when counts of insects, plants, animals, or inventory are made, but it is more usual for measurements to be able to take on all values between two limits and not be confined only to whole numbers. For example, all heights between four feet and eight feet are possible because men's heights do not increase by jumps of exactly one inch with nothing in between. Measurements such as this are said to be continuous.

Grouping was introduced in the first chapter as a method of comparing the relative frequencies of various measurements. In order to make the comparison worthwhile it was necessary to choose the grouping interval sufficiently large to include several measurements. As we increase the size of the sample, however, the grouping interval can be made smaller. In fact for a large enough sample, the grouping interval can be made so small that the blocks of the histogram blend together and appear as a curve. That is, the theoretical frequency curve results from the process of indefinitely increasing the same size and thereby reducing the grouping interval indefinitely.

The important point that we want to make is the following. The theoretical frequency curve retains the property of the histogram in that the area under the curve between any two values gives the frequency between the two values. Usually the histogram is plotted in terms of relative frequencies. Then the theoretical frequency curve is such that the area between any two values gives the probability that a random draw will occur between those two values. Two terms often used are random variable and the probability density of the random variable. The numerical value of the random draw represents the random variable, and the theoretical (relative) frequency curve represents its probability density.

Exercises

1. In each of the following examples, say whether the population is existent or conceptual: all ball-bearings that can be turned out by a given machine; all the whales in the sea; all wheat plants; all bridge hands in card playing; all the coal reserves of the world; all people that

like music; the weather in Boston; the flight of a wild goose; next year's orange crop.

2. In each of the following examples, say whether the population is finite or infinite: all poker hands in card playing; all hydrogen atoms; all the stars; all people; automobiles.

3. Toss a die 10 times and draw the histogram; toss the die 90 more times and draw the revised histogram.

4. Discuss methods of obtaining random samples (with replacement) from the following populations, and give for each a possible measurement that might be made on the members in the sample: (a) Apples on a tree, (b) People in a state, (c) Students in a university, (d) A month's production of automobile tires, (e) Fish in a lake, (f) Seeds planted in a field.

5. Suppose that we wish to determine the average number of people in a family in Boston. Would the recording of the number of people in each student's family in a random selection of students at the University of Massachusetts in Boston be a reasonable way to obtain the necessary data?

6. What possible bias could result from a sample obtained by selecting every tenth item in a population? Why is this not a random sample?

7. Suppose we ask whether two school classes are significantly different, based on the results of some tests. In what populations are we interested?

8. Suppose there are 10,000 students in attendance at a given university, and a test has been given to 400 of these students. The complete set of 10,000 students we could have tested is called a

   (a) mean
   (b) sample

(c) population

(d) statistic

9. The statement "Mary's dancing is above average" means Mary's dancing

(a) represents some middle position or value

(b) is mutually exclusive

(c) on some scale is above some middle position or value

(d) is the mode.

10. Let k represent any one of the set of numbers 1, 2, 3, 4, 5, 6, 7, 8, 9. The subset $3 < k < 6$ is

(a) 3, 4, 5, 6

(b) 3, 4, 5

(c) 4, 5, 6

(d) 4, 5

11. Which of the following variables would you regard as continuous?

(a) number of students in each class at a university

(b) number of children per married couple

(c) time spent in preparing an assignment

(d) number of correct responses on a test

12. Which of the following variables would you regard as discrete?

(a) reaction time to stop a car

(b) annual income of professors at a university

(c) speed of running a maze

(d) time spent waiting for an elevator

13. Which word does not belong with the others below?

(a) data point

(b) statistic

(c) sample value

(d) parameter

14. When tossing a perfect die each face is

(a) equally likely

(b) biased

(c) of zero probability

(d) of probability one

15. The population mean μ can be found by means of a sample provided the sample size

(a) is greater than 1000

(b) exhausts the whole population

(c) is finite

(d) in the 95% range

16. From sample to sample, the theoretical frequency distribution is

(a) dependent upon sample size

(b) the same

(c) the same as the empirical frequency distributions

(d) the most variable of the empirical frequency distributions

17. For a random sample with replacements, each sample member is

(a) except for the first, dependent on the previous sample member

(b) dependent upon all of the other sample members

(c) equally likely and independent

(d) equal to every other sample member

18. In order to keep a sample random, an extreme data point ought to be

(a) thrown out
(b) retained
(c) reduced
(d) emphasized

Chapter 11. Normal distribution

"How poor are they that have not patience? What wound did ever heal but by degrees? (*Othello*).

Background
One of the most important theoretical frequency distributions in statistics is the *normal distribution.* Many statistical procedures are based on knowledge of this distribution. Unlike most other important geometrical forms, the shape of the normal curve does not appear in any common object familiar to the eye. The closest object that resembles a normal curve is the outline of a bell. The normal curve is often described as a symmetrical bell-shaped curve, extending infinitely far in both positive and negative directions.

The concept of the normal distribution is one that has universal appeal. Nearly all scientists believe in the normal distribution: Observers because they believe it is a theorem of mathematics; mathematicians because they believe that it is aesthetically satisfying; aestheticians because they believe it is philosophically true; philosophers because they believe it is a fact of observation. A few neither believe it nor disbelieve it; these maintain that the normal distribution is a convenient statistical procedure for the bookkeeping of data either in raw form or under suitable transformations, and about bookkeeping procedures it ought to be asked, not if they are true or false, but are they useful.

The central limit theorem and the normal curve
It can be shown mathematically that whenever a measurement is the sum of a large number of small independent effects, no one of which predominates, the distribution of the measurement will take approximately the same general bell-like shape. This bell-like shape is the normal curve, and the knowledge of its central value (mean) and spread (standard deviation) determines the curve completely. That is,

once the mean and standard deviation are known all else is determined. The mean and standard deviation are called *parameters,* so the normal curve is determined by two parameters. Since the normal distribution is a theoretical frequency curve, it is customary to use Greek letters to represent its parameters. More specifically, the Greek letter μ (called mu) denotes its mean, and the Greek letter σ (called sigma) denotes its standard deviation. The figure illustrates the normal curve with mean μ and standard deviation σ.

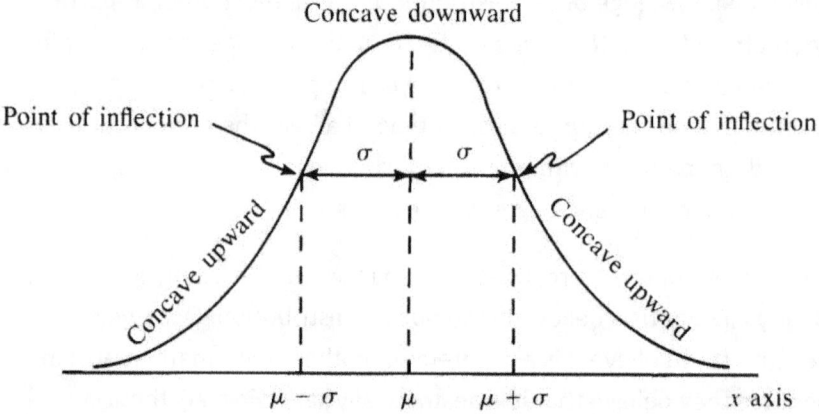

The normal curve is symmetrical on each side of the mean μ. The total area between this curve and the x-axis is one square unit. The curve is concave downward for x within $-\sigma$ to σ of the mean, and concave upward for x outside of this range. As a result there are two points of inflection, one at $\mu - \sigma$ and the other at $\mu - \sigma$. (A point of inflection is the point where the slope stops getting steeper and starts getting flatter as you slide down a hill.) Thus the standard deviation is the physical horizontal distance from the mean to either point of inflection. (Just as all circles have the same general shape and are determined by the center and radius, all normal curves have the same general shape and are determined by the mean and standard deviation. The center of the circle and the mean of the normal curve are location parameters, whereas the radius of the circle and the standard deviation of the normal curve are spread parameters. As in its class the circle is conceptually the simplest possible geometric shape, so also in its class the normal curve is the simplest possible geometric shape. One might say that the normal curve plays a role in statistics that is analogous to the role played by the circle in ancient mathematics and astronomy.)

As we have just stated, the normal distribution results (in the limit) in the case of any variable made up as a sum of many small independent variables as long as each small independent variable has a negligible effect on the sum. That is, no one of the variables in the sum is allowed to have a predominant effect on the sum. This remarkable result is called the *central limit theorem.*

Let us now look at some examples of the central limit theorem. The total consumption of electric power delivered by the Edison Company is the sum of the quantities consumed by the various customers, so the total consumption has a normal distribution. Likewise we would expect a normal distribution for the total gain or loss on the risk business of an insurance company, as the total is the sum of the gains or losses on each single policy. We would expect the total error committed in a physical or astronomical measurement to have a normal distribution, as the total error is the sum of a large number of mutually independent elementary errors. The yield of wheat on a farm would be normally distributed, as it is the result of many small effects due to environmental factors such as rainfall, temperature, soil fertility, pest activity, and plant competition and to a multitude of genetic factors. The distribution of the result of 10 tosses of a coin is approximately normal, because the total score is the sum of ten independent scores resulting from each toss.

Despite this seeming universality of the normal distribution, we cannot expect it to apply to every measurement for various reasons. First, there may not be a large number of small effects. Second, one particular effect may predominate. Third, the effects may not be independent, as with rainfall and temperature, although a small degree of dependence may not seriously affect the normality provided there are a large enough number of other independent effects. Fourth, the effects may not be additive.

The standard normal distribution

There are an infinite number of normal distributions. However all normal distributions have the same general shape, namely the shape of a bell, and they differ from each other only with respect to their means and standard deviations. A particular normal distribution is completely determined by the specification of its two parameters: its mean μ and its standard deviation o-. We say that the normal distribution represents a two parameter family of curves and both parameters must be assigned to specify a particular normal curve. If the standard deviation o-is fixed and if we vary the mean it, we have a family of curves with identical shape but with different locations on the horizontal axis. If the mean μ is fixed and if we vary the standard deviation a-, we have a family of curves with identical location on the horizontal axis but with different spreads.

One of the main uses of the normal curve is to calculate probabilities. For this purpose we direct our attention on one particular normal distribution whose mean is 0 and whose standard deviation is one. This distribution is called the *standard normal distribution*. The abscissa for the standard normal distribution is labeled z, and z is called the *standard normal variable*. Probabilities are assigned to intervals of possible values of the variable z, rather than to single values of z. *The probability that on a single draw the variable z will fall within a stated interval is equal to the area under the curve within that interval.* For example, the area under the curve within the interval $-1.96 < z < 1.96$ is 0.95. Thus the probability that a value of z drawn at random from the population will fall in the interval $-1.96 < z < 1.96$ is 0.95. We illustrate this result by the figure

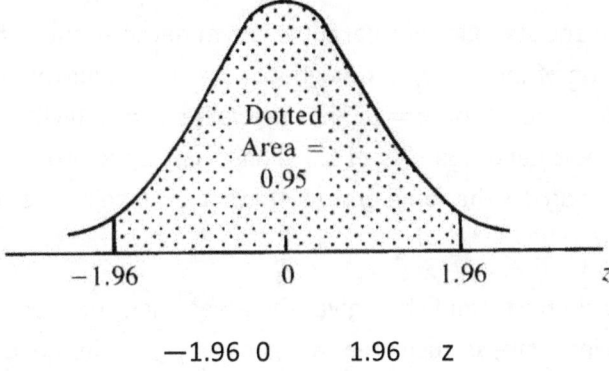

−1.96 0   1.96   z

or by the equation

$$P(-1.96 < z < 1.96) = 0.95$$

which we read as "The probability of z between −1.96 and 1.96 is 0.95."

The probability that an observed z lies between $-\infty$ and ∞ is one. We can write this statement as

$$P(-\infty < z < \infty) = 1.00$$

which states that the area under the standard normal curve is equal to one. The essential thing to remember is: The probability of z falling into any given interval is equal to the area under the curve for that interval.

The standard normal distribution is the curve shown in the figure

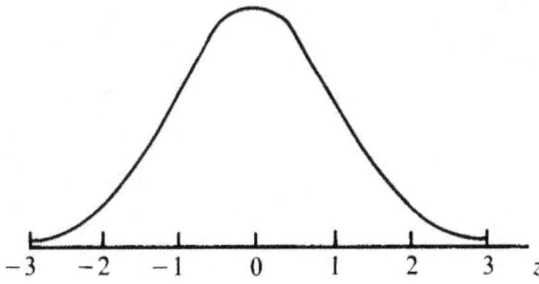

We see that the standard normal curve is symmetric about $z = 0$. That is, the portion of the curve to the right of $z = 0$ is the mirror image of the portion to the left of $z = 0$. Thus the value $z = 0$ divides the curve and the total probability of 1.0 in half. Hence the probability that z will be less than 0 is the same probability that z will be greater than 0, namely 0.5.

There is no simple formula that gives the areas under the normal curve for various intervals, so such areas are generally given in the form of tables. A table is given in the Appendix (page 180). However, before one turns to tables one can get a good idea of these probabilities by memorizing certain key values.

We divide the z-axis into 8 key intervals. The probability assigned to each of these intervals is as follows:

Interval	Probability
$-\infty$ to-2	0.02
-2 to -1	0.14
-1 to $-1/2$	0.15
$-1/2$ to	0.19
0 to 1/2	0.19
1/2 to 1	0.15
1 to 2	0.14
2 to 00	0.02
Total $-\infty$ to ∞	1.00

This table may be shown graphically as

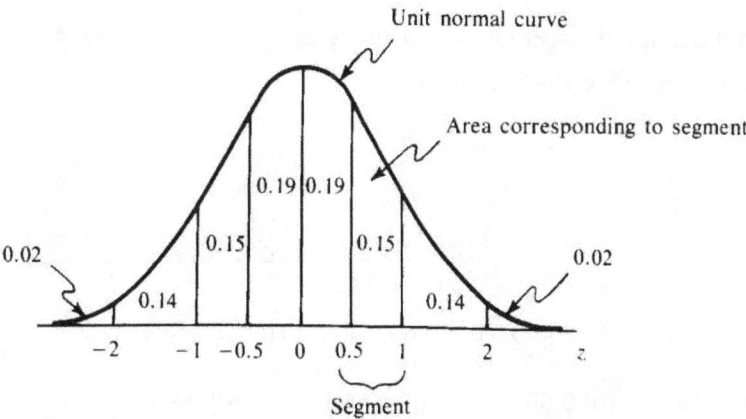

Unit normal curve

From this table or picture we can build up various probabilities. Certain relationships hold among areas and hence among probabilities. As long as areas don't overlap we can add them. That is, as long as intervals don't overlap, we can add the corresponding probabilities.

Exercises

1. The probability of z being less than 0.5 is

   (a) $P(0.5 < z < 1.0)$
   (b) $P(-1.0 < z < 0.5)$
   (c) $P(-\infty < z < 0.5)$
   (d) $(0.5 < z < \infty)$

2. For a standard normal distribution, the probability that z will occur below 2 is

   (a) 0.48   (b) 0.99   (c) 0.98   (d) 0.49

3. A particular normal distribution is determined by the values of its

   (a) mode and mean
   (b) mean and standard deviation
   (c) median and mode
   (d) median and mean

4. On the unit normal curve $P(-1.0 < z < 0) + P(0 < z < 0.5)$ corresponds to the shaded area in

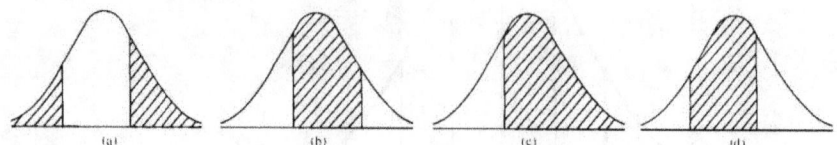

(a)   (b)   (c)   (d)

5. Given $P(-\infty < z < 0.5) = 0.69$, then $P(0.5 < z < \infty)$ is

(a) 0.19   (b) 0.69   (c) 0.31   (d) 0.50

6. Which is not correct for the standard normal distribution?

(a) The total area under the curve is one.
(b) The mean occurs at z = 0.
(c) The probability of any given value of z is 0.5.
(d) The curve is symmetric about the mean.

7. Since the standard normal curve is symmetrical about the mean $\mu = 0$, which would be correct?

(a) $P(-\infty < z < \infty) = 0.5$
(b) $P(0 < z < \infty) = 1.0$
(c) $P(-\infty < z < 0) = P(0 < z < \infty)$
(d) None of these

8. For a standard normal curve, the probability that z will fall below a given value K plus the probability that z will fall above the given value K is

(a) $P(z < K) + P(z < \infty)$
(b) $P(z > K) + P(z > 0)$
(c) one
(d) 0.50

9. For a standard normal distribution which of the following is correct

(a) $P(-\infty < z < -1) = 0.84$
(b) $P(-\infty < z < 1) = 0.84$
(c) $P(-1 < z < 1) = 0.84$
(d) $P(1 < z < \infty) = 0.84$

10. For a standard normal distribution which is correct?

(a) $P(-1 < z < \infty) = 0.16$
(b) $P(< z < 2) = 0.02$
(c) $P(-\infty < z < -2) = 0.98$
(d) $P(-\infty < z < -2) = 0.02$

11. Using the figure given for the standard normal curve, find the:

(a) Probability that a z-score will be greater than 1
(b) Probability that a z-score will be between 0.5 and 2
(c) Probability that a z-score will be greater than -0.5
(d) Probability that a z-score will be between -1 and 2
(e) Find the z-score such that the probability of a larger value is 0.02
(f) Find the z-score such that the probability of a larger value is 0.84
(g) Suppose that the z-score lies between $-b$ and $-b$. Find the value of b such that the probability is 0.96

12. Given the correspondence between x and z given by

$$z = (x - \mu)/\sigma$$

Suppose the mean $\mu = 75$ and the standard deviation $\sigma = 10$ for a normal population. What percentage of all scores lies between 70 and 80? Hint. Convert x scores to z scores. Then find area of segment between the two z scores.

13. The mean $\bar{x}$ of a sample of size n has a sampling distribution with mean μ and variance σ^2/n. The standardized mean is

$$z = \frac{x - \mu}{\sigma} \sqrt{n}$$

Given the normal distribution with $\mu = 75$ and $\sigma = 10$, what is the probability that the mean z of a sample of $n = 25$ will differ from the population mean μ by less than one?

14. The lengths of adult snakes of a certain species are found to have a normal distribution with mean 14 inches and standard deviation 2 inches.

(a) Sketch the normal curve.
(b) What fraction of these snakes are over 16 in. long? under 12 in? between 12 and 16 in.?
(c) What fraction of these snakes are between 14.4 in. and 18 in.? between 12.6 and 15 in.? between 12 and 12.6 in.?

Solution:

(a) To sketch this bell-shaped curve, first mark the mean and standard deviation on the x-axis, then find the height of the bell $0.4/\sigma = 0.4/2 = 0.2$ and plot this point above the mean. Next mark points over the standard deviation about 3/5 the height of the bell, and then draw in a bell-shaped curve through these points.

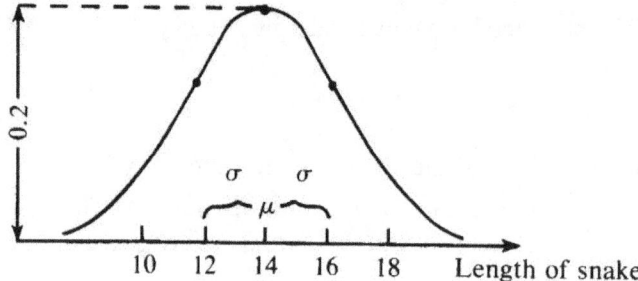

(b) 16 inches is 2 inches over the mean, or one standard deviation above the mean. You should remember that 0.16 (or approximately 1/6) of cases lie more than one standard deviation above the mean for a normal curve. Similarly 12 inches is one standard deviation below the mean, and 0.16 of cases occur below this level. Also 0.68 of cases lie within one standard deviation of the mean, i.e. between 12 and 16 inches.

(c) To answer these questions, you need the Table for the Normal Distribution. It is a good idea to shade in the area required for each question. The fraction *between* 14.4 and 18 inches is shown by the shaded area

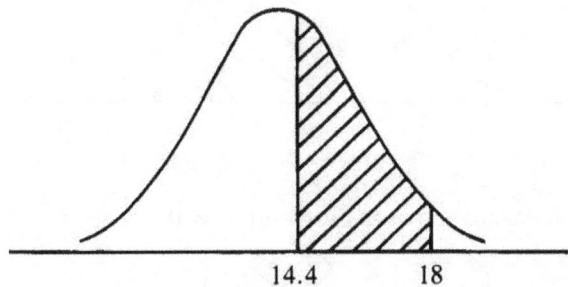

14.4   18

One can already guess that answer to be about 0.4, since a little less than half the area under the curve is shaded in. Now calculate the z-values, the number of standard deviations out from the mean for 14.4 inches and 18 inches. We see that 14.4 inches is 0.4 inches above the mean or $z = 0.4$ inch/2 inches $= 0.2$ standard deviations above the mean. Similarly 18 inches is $z = 2$ standard deviations above the mean. The Table gives the area between $z = 0$ (the mean) and each value of z in the table: for $z = 0.2$, the area of 0.0793 given in the table represents this area:

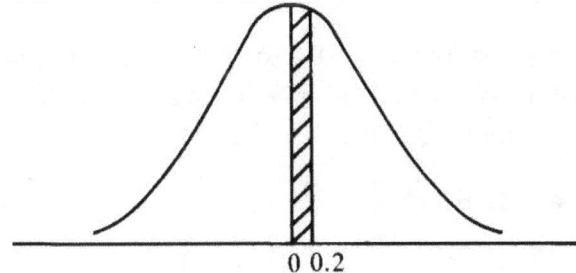

0 0.2

while for $z = 2$, the value of 0.4772 represents this area:

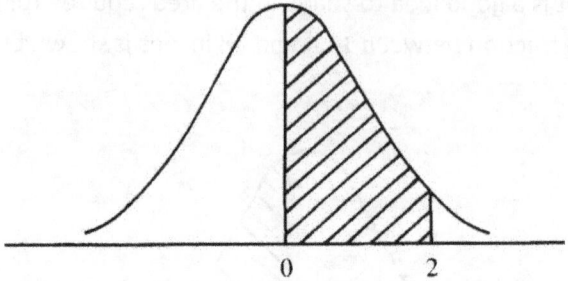

We want the area between $z = 0.2$ and $z = 2$

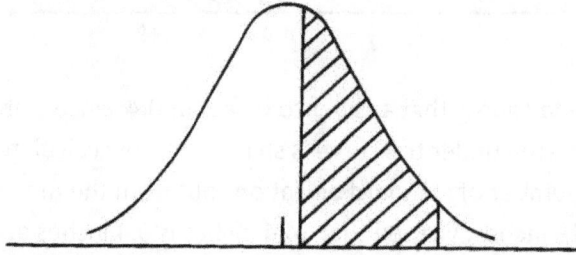

which is clearly the second figure *minus* the first figure. Answer =

0.4772 - 0.0793 = 0.3979 of snakes have lengths between 14.4 inches and 18 inches.

Between 12.6 and 15 inches:

Total area = (Area from $z = -0.7$ to 0) + (Area from $z = 0$ to 0.5)
= (Area from $z = 0$ to 0.7) + (Area from $z = 0$ to 0.5)
= 0.1915 + 0.2580 = 0.4495

Between 12 and 12.6 inches:

Total area = (Area from $z = -1$ to -0.7)
= (Area from $z = 0.7$ to 1)
= (Area from $z = 0$ to 1) − (Area from $z = 0$ to 0.7)
= 0.3413 − 0.2580 = 0.8333

15. Company X makes pound-boxes of butter, but variations (errors) naturally occur from box to box. Suppose to be on the safe side, the

management sets the machine to produce boxes of average net weight a little above 1 lb., say at 1.01 lb. The variability makes the distribution of new weights in the boxes a normal curve with mean 1.01 lb. and standard deviation 0.02 lb.

(a) Sketch the density function and give its formula.

(b) What fraction of boxes exceed 1.03 lb. in weight? are less than 0.99 lb. in weight? are between 0.99 and 1.03 lb.?

(c) What fraction of boxes are underweight, i.e., less than 1.00 lb.?

16. Company Y is a shoddy outfit also making pound boxes of butter. Its machines are less accurate, so that the standard deviation is 0.05 lb. To make matters worse, the management cheats a little and sets the machine to average only 0.98 lb.

(a) Sketch the density function, assumed normal.

(b) 0.16 of the boxes exceed _____ in net weight, while 0.16 of the boxes are below _____ in weight, and the remaining 0.68 are between _____ and _____ in weight.

(c) What is he fraction of the boxes that are underweight?

17. The College Entrance Examination Board test scores are scaled to approximate a normal distribution with mean $\mu = 500$ and standard deviation $\sigma = 100$.

(a) Sketch the density function and give its formula

(b) 0.16 of the students *taking the exam* score over ____ while 0.16 of the students score below ____ , and the remaining 0.68 of students score between ____ and ____

(c) What fraction of students taking the exam score between 450 and 650? over 650?

18. Which of the following would have an approximately normal distribution? If possible, sketch the distribution (normal or non-normal)

(a) adult male's heights

(b) the distribution of family incomes in the U.S.

(c) heights of all people in the U.S.

(d) estimates of the length of an object by a large group of people all using the same ruler in turn

(e) daily production of a certain factory

19. The area under the normal curve is

(a) equal to 0.95

(b) between plus and minus 2 standard deviations

(c) one

(d) one hundred

20. For a normal distribution with a mean of 60 and a standard deviation of 10, what is the chance that a score is greater than 60?

(a) 0.25   (b) 0.196

(c) 0.50   (d) 70

Chapter 12. Expectation

"We know what we are but know not what we may be." (*Hamlet*)

Properties of expectation

One of the most useful concepts in statistics is that of expected value or expectation, denoted by the symbol *E*. In words, the symbol *E* is an operator that stands for any of the equivalent expressions given here:

Expectation of
Expected value of
True mean of
Population mean of
E of

For example, instead of saying that μ is the population mean of the score x, we can merely write

$$\mu = Ex$$

which in words is μ is equal to E of x. The expected value can always be thought of a population average of the quantity that follows the symbol E. The symbol E is a convenient and useful shorthand notation for a concept which is used over and over again in statistics. Usually we use the symbol μ alone, but to explicitly show that /1 is the mean of x we can attach x as a subscript to μ. Thus,

$$\mu_x = Ex$$

Now let us develop some useful properties connected with the expected values of different kinds of scores. The simplest kind of score is one capable of taking on only a single value c. Such a score that doesn't change at all, of course, is a constant. For example $c = 100$ would correspond to the case of giving everyone in a class 100 percent regardless of his performance on the examination. Since the score will always equal $c,$ it will clearly be equal to c on the average, so its expected value must be $c.$ We express this property as

$$Ec = c$$

and say that the expected value of a constant score c is equal to $c.$

Next let x be any score, and c be a constant. We can obtain a new score by adding c to the value of $x.$ For example we might add 50 points to each score obtained on an examination. The new score would be given by $x + c.$ The expectation of the new score $x + c$ is given by

$$E(x + c) = Ex + c$$

and we say that the expected value of the new score $x + c$ is equal to the expected value of the old score x plus the constant $c.$ In words, when a constant c is added to a score, the same constant c is added to its expectation. For example, if each score x is increased by 50 points, then the class average will necessarily be increased by 50 points also.

Let us next find out what happens when we multiply each score x by a constant $c.$ For example, suppose we multiply each score by 10. We would find that the new class average is then 10 times that of the old class average. In symbols we have

$$E(cx) = c\,Ex$$

which in words is that the expected value of cx is equal to c times the expected value of $x.$

Finally we come to one of the most useful properties of expectation, namely the addition property. Let each student obtain two scores x and $y,$ and let us obtain a new score by adding them together. The new score is then the sum $x + y.$ The class average of the new score is then equal to the sum of the class averages of the old scores. That is,

$$E(x + y) = Ex + Ey$$

or, in words, the expected value of the sum of two scores is the sum of the expected values of each score.

It is easy to extend this addition property to more than two kinds of scores. For example, each student in a class may receive a score x in math, y in English, and z in science. The students total score in the three subjects is $x + y + z$. The class average of the total score $x + y + z$ is the sum of the averages Ex, Ey, Ez of the individual subjects, that is

$$E(x + y + z) = Ex + Ey + Ez$$

More generally, if $x_1, x_2, \cdots, x_n$ are any n types of scores, then the expected value of their sum is equal to the sum of their expected values, which in symbols is

$$E(x_1 + x_2 + \cdots + x_n) = Ex_1 + Ex_2 + \cdots + Ex_n$$

Variance

The expected value of a score x is equal to the true mean (or population mean) μ of the score, that is $Ex = \mu$. We recall that the population variance of the score x is a true measure (or population measure) of the extent to which the population distribution of scores is dispersed away from the true mean. More specifically, the population variance of the score x is the expected value of $(x - \mu)^2$; that is,

$$\text{var } x = E(x - \mu)^2$$

In words, the population variance of a score is the expectation of the square of the difference between the score and its expectation. Before, we have introduced the symbol cr^2 for the population variance, so

$$\sigma^2 = \text{var } x = E(x - \mu)^2$$

However the symbol $\text{var } x$ is more explicit, as it clearly shows that we are dealing with the score x.

Variance has properties analogous to the properties of expectation. The first property is that for any constant c, we have

$$\text{var } c = 0$$

The reason for this property is as follows. If a score is always equal to a constant, then the constant will be the expectation of the score. That is, if $x = c$ then $\mu = c$ also. Therefore the score never differs from its expectation, and hence the difference $x - \mu$ is always $c - c$ or 0. The square of zero is zero, and so is its expectation.

The next property is that for any constant c

$$\text{var }(x + c) = \text{var } x$$

We can obtain this result as follows. If we add the constant c to the score x, the expectation of x is also increased by c. It follows that the difference $x - \mu$ is unchanged, so that the variance is also unchanged.

If c is any constant, then

$$\text{var}(cx) = c^2 \text{ var } x$$

We justify this property as follows. Multiplying the score x by c will also multiply its expectation μ by c. Hence the difference is $cx - c\mu = c(x - \mu)$. That is, the new difference is c times as large as before. The square of the new difference is $c^2(x - \mu)^2$ which is c^2 times as large as before. It follows that

$$\text{var } cx = E[c^2(x - \mu)^2] = c^2 E(x - \mu)^2 = c^2 \text{var } x$$

That is, the variance of cx is c^2 times the variance of x. In the special case when $c = -1$, we have

$$\text{var }(-x) = (-1)^2 \text{var } x = \text{var } x$$

This property shows that multiplication of a score by a constant results in the multiplication of the variance of the score by the square of the constant. This is because the variance is the expectation of a squared quantity. If we convert from feet to inches, each score is multiplied by 12, but the variance is multiplied by $(12)^2 = 144$. As a result it is usually more convenient to use a spread parameter that changes by the same scale factor, rather than by the square factor. As we have seen, such a parameter is the standard deviation σ, defined as the positive square root of the variance. Usually we use the symbol σ alone, but to explicitly show that σ is the standard deviation of the score x we can attach x to σ as a subscript. Thus

$$\sigma_x = \sqrt{\text{var } x}$$

or equivalently

$$\sigma_x^2 = \text{var } x$$

The foregoing property becomes

$$\sigma_{cx} = c\sigma_x$$

which says that the standard deviation of a constant times a score is equal to the constant times the standard deviation of the score. This property follows from

$$\sigma_{cx} = \sqrt{\text{var } cx} = \sqrt{c^2 \text{var } x} = c\sqrt{\text{var } x} = c\sigma_x$$

Thus the standard deviation of a score in inches is just 12 times the standard deviation of the score in feet.

We now come to the addition property. By analogy with the addition property of expectations, we would hope that there would be an addition property for variances that would say that the variance of the sum of two scores is the sum of their variances. Such an addition property is true provided that the two scores are *uncorrelated.* We will take up correlation later, but for the moment we need to know only that independent scores are always uncorrelated. Thus for two independent scores x and y we have the addition property for variances given by

$$\text{var}(x + y) = \text{var } x + \text{var } y$$

In words, the variance of the sum of two independent scores x and y is equal to the sum of the variance of x and the variance of y. More generally, the variance of the sum of n independent scores is equal to the sum of their variances:

$$\text{var } (x_1 + x_2 + \cdots + x_n) = \text{var } x_1 + \text{var } x_2 + \cdots + \text{var } x_n$$

Random sample

We recall that a random sample of size n is a set of n independent draws $x_1, x_2, \dots, x_n$ from a given population. If the population mean is μ and the population variance is σ^2, then each of these draws has μ as its expected value and σ^2 as its variance. That is,

$Ex_1 = \mu$	$var\ x_1 = \sigma^2$
$Ex_2 = \mu$	$var\ x_2 = \sigma^2$
$Ex_n = \mu$	$var\ x_n = \sigma^2$

The sample mean is

$$\bar{x} = \frac{x_1 + x_2 + \cdots + x_n}{n}$$

which we can write as

$$\bar{x} = \frac{\sum x}{n}$$

The expected value of the sample mean is

$$E\bar{x} = \frac{E \sum x}{n} = \frac{1}{n} E(x_1 + x_2 + \cdots + x_n)$$

$$= \frac{1}{n}(Ex_1 + Ex_2 + \cdots + Ex_n) = \frac{1}{n}(\mu + \mu + \cdots + \mu) = \frac{n\mu}{n}$$

which finally gives

$$E\bar{x} = \mu$$

In words, the *expected value of the sample mean is equal to the population mean.*

The variance of the sample mean is

$$\text{var } \bar{x} = \text{var}\left(\frac{\sum x}{n}\right) = \frac{1}{n^2}\text{var}\left(\sum x\right)$$

$$= \frac{1}{n^2}\text{var}(x_1 + x_2 + \cdots + x_n) = \frac{1}{n^2}(\text{var } x_1 + \text{var } x_2 + \cdots + \text{var } x_n)$$

$$= \frac{1}{n^2}(\sigma^2 + \sigma^2 + \cdots + \sigma^2) = \frac{n\sigma^2}{n^2}$$

which finally gives

$$\text{var } \bar{x} = \frac{\sigma^2}{n}$$

In words, the *variance of the sample mean is equal to the population variance divided by the sample size.* Taking square roots we have

$$\sigma_{\bar{x}} = \frac{\sigma}{n}$$

In words, the *standard deviation of the sample mean is equal to the population standard deviation divided by the square root of the sample size.*

Sample variance

We recall that the sample variance is defined as

$$s^2 = \frac{1}{n-1}\sum(x - \bar{x})^2$$

If we define a new score y as $y = x - \mu$, then clearly

$$x - \bar{x} = y - \bar{y}$$

so in terms of the new score the sample variance is

$$s^2 = \frac{1}{n-1} \sum (y - \bar{y})^2$$

That is, we can substitute y for x, and $\bar{y}$ for $\bar{x}$, in the formula for s^2
We also recall the alternate formula for the sample variance given by
(see page 10)

$$s^2 = \frac{1}{n-1} \left(\sum x^2 - n\bar{x}^2 \right)$$

Because we can substitute y for x, and $\bar{y}$ for $\bar{x}$, we obtain the
equivalent formula

$$s^2 = \frac{1}{n-1} \left(\sum y^2 - n\bar{y}^2 \right)$$

Taking expectations, we have

$$Es^2 = \frac{1}{n-1} E \left(\sum y^2 - n\bar{y}^2 \right)$$

which is

$$Es^2 = \frac{1}{n-1} \left(E \sum y^2 - n E\bar{y}^2 \right)$$

Let us now look at the terms in the parenthesis on the right. The first
term can be written as

$$E \sum y^2 = E(y_1^2 + y_2^2 + \cdots + y_n^2)$$
$$= Ey_1^2 + Ey_2^2 + \cdots + Ey_n^2$$
$$= E(x_1 - \mu)^2 + E(x_2 - \mu)^2 + \cdots + E(x_n - \mu)^2$$
$$= \text{var } x_1 + \text{var } x_2 + \cdots + \text{var } x_n = \sigma^2 + \sigma^2 + \cdots + \sigma^2 = n\sigma^2$$

Here we have used the fact that $y = x - \mu$. Thus the second term is

$$-n E\bar{y}^2 = -n E(\bar{x} - \mu)^2 = -n \text{ var } \bar{x}$$

Here we have used the fact that $\bar{y} = \bar{x} - \mu$. From the last section we know

$$\text{var}\,\bar{x} = \frac{\sigma^2}{n}$$

Thus the second term is

$$-n\,E\bar{y}^2 = -n\,\frac{\sigma^2}{n} = -n\sigma^2$$

The equation for Es^2 becomes

$$Es^2 = \frac{1}{n-1}(n\sigma^2 - \sigma^2) = \frac{1}{n-1}(n-1)\sigma^2 = \sigma^2$$

In words, the *expectation of the sample variance is equal to the population variance.*

Unbiased estimates

We recall that a statistic is a quantity computed from the data. For example the sample mean $\bar{x}$ is found by summing the sample observations and then dividing the sum by the number n of observations. The sample mean $\bar{x}$ can be used as an estimate of the population mean. Because the expected value of $\bar{x}$ is equal to μ, we say that i is an unbiased estimate of μ.

Generally, we say that an *estimate is unbiased if its expected value is equal to the population parameter that we are trying to estimate.*

As another example we compute the statistic s^2 to estimate the population variance σ^2. As we have seen in the preceding section, the expected value of s^2 is equal to σ^2. Thus s^2 is an unbiased estimate of the population variance σ^2. We recall that s^2 is found by summing the squared deviations of x from the sample mean $\bar{x}$ and then dividing the result by $n-1$. If instead we had divided by n, we would not have obtained an unbiased estimate.

Probability for Business and Economics

In some cases we may know the population mean μ. In these cases we can form the estimate of the population variance given by

$$\frac{1}{n}\sum(x-\mu)^2$$

Because

$$E\left[\frac{1}{n}\sum(x-\mu)^2\right]=\frac{1}{n}\sum E(x-\mu)^2$$
$$=\frac{1}{n}\sum\sigma^2=\frac{1}{n}(\sigma^2+\sigma^2+\cdots+\sigma^2)=\frac{1}{n}(n\sigma^2)=\sigma^2$$

we see that this estimate is unbiased.

The important points to remember are these: If the deviations are measured about the sample mean $\bar{x}$, then we divide the sum of squared deviations by $n-1$ in order to obtain an unbiased estimate of σ^2. On the other hand if the deviations are measured about the population mean μ, then we divide the sum of squared deviations by n in order to obtain an unbiased estimate of σ^2.

Exercises

1. Let the fixed monthly salary of a salesman be 500, and let x be the random amount he earns through commissions during a month. If the expected value of his monthly commissions is 750, what is the expected monthly income?

2. If a gambler on straight bets expects to earn 100, and on side bets 50, what are his expected total earnings?

3. Because a class did poorly on an exam, the teacher doubled everybody's grade. If the old class average is 37, what is the new class average?

4. Show that $E(x-y)=Ex-Ey$. Hint: $x-y=x+(-1)y$.

5. If c and d are constants, show that $E(cx+dy)=cEx+dEy$.

6. Suppose that each of the three scores x, y, z has the same expectation μ. Show that $E(x + y + x) = 3\mu$.

7. Suppose that each of the n scores x_1, x_2, ... , x_n has the same expectation μ. Let $\bar{x}$ be the arithmetic mean $(x_1 + x_2 + ... + x_n)/n$. Show that $E\bar{x} = \mu$.

8. If the physical education scores x have variance 25 and the science scores y have variance 16 (and the scores are independent) what is the variance of the score $x + y$.

9. If a score is multiplied by a negative constant, then is the resulting standard deviation negative?

10. If x and y are independent, x has a standard deviation of 6 and y has a standard deviation of 8, then what is the standard deviation of their sum?

11. If x and y are independent and each has the same standard deviation σ, then what is the standard deviation of the sum $x + y$?

12. If x and y are independent and c and d are constants, show that

$$\text{var}\,(cx + dy) = c \text{ var } x + d \text{ var } y$$

13. If x and y are independent, show that

$$\text{var}\,(x - y) = \text{var } x + \text{var } y$$

14. Suppose that each of the three independent scores x, y, z has the same variance σ^2. Show that

$$\text{var}\,(x + y + z) = 3\sigma^2$$

15. Suppose that each of the n independent scores $x_1, x_2, \ldots, x_n$ has the same variance σ^2. Let $\bar{x}$ be their arithmetic mean $(x_1 + x_2 + \cdots + x_n)/n$. Show that $\text{var}\,\bar{x} = \sigma^2/n$. $x_1 + x_2 + \cdots + x_n$

16. If the expected value of x is 5 then the expected value of $-x$ is (a) 0.0 (b) 0.5 (c) 0.25 (d) –5

17. If the variance of x is 5 then the variance of $-x$ is (a) –5 (b) $\sqrt{5}$ (c) 5 (d) 25

18 If the expected value of x is 5 and the expected value of y is 10 then the expected value of $x - y$ is

(a) 15 (b) 10 (c) 5 (d) –5

19. If the standard deviation of x is 3, if the standard deviation of y is 4, 2. and if x and y are independent, then the standard deviation of $x + y$ is

(a) 3 (b) 4 (c) 5 (d) 7

20. If the expected value of x is 50 then the expected value of $\bar{x}$ is

(a) 0 (b) 50 (c) $50n$ (d) $50/\sqrt{n}$

21. If the standard deviation of x is 5 then the standard deviation of x is

(a) 5 (b) $5n$ (c) $5/\sqrt{n}$ (d) 5

n

22. If the standard deviation of x is 5 then the standard deviation of 10x is (a) 5 (b) 25 (c) 50 (d) 250

23. If the variance of x is one, then the variance of 10x is (a) 1 (b) 10 (c) 100 (d)

24. If x and y are independent, and if x has standard deviation 4 and y has standard deviation 3, the x — y has standard deviation

(a) 1 (b) V 7 (c) 5 (d) 7

25. If the variance of x is cr^2 then the standard deviation of z is (a) $(7^2/n$
(b) *not* (c) *cr n* (d) o/n

26. Given the sets of numbers 2, 5, 8, 11, 14 and 2, 8, 14, find (a) the mean of each set, (b) the variance of each set, (c) the mean of the combined or "pooled" sets, (d) the variance of the combined or pooled sets.

27. A student received a grade of 84 on a final examination in mathematics for which the mean grade was 76 and the standard deviation was 10. On the final examination in physics for which the mean grade was 82 and the standard deviation was 16, he received a grade of 90. In which subject was his relative standing higher

CHAPTER 13 Decision making

"But, soft! what light through yonder window breaks?
It is the east, and Juliet is the sun." (*Romeo and Juliet*

Null hypothesis, significance, and the level of significance
Statistics may be described as the science of making decisions from observations. Often we are faced with a decision in which the two alternatives are so equally balanced that we cannot make up our minds. For hours we will sit pondering first the one and then the other and all the time growing steadily unhappier. Indeed we become uncomfortably aware of the resemblance of our case to that of "Buridan's donkey." The reference here is to the French philosopher, Jean Buridan, in the fourteenth century, who stated the following: "If a hungry donkey were placed exactly between two haystacks equal in every respect, the donkey would starve to death, because there would be no motive why it should go to one rather than to the other haystack."

What Buridan did not take into account was statistical theory, for his statement does not recognize the existence of a random factor. The donkey is bound to turn his head in a random way so that one haystack comes into a better view, or to move his feet randomly so that one haystack becomes closer, and in the end he would be eating from the haystack seen better or closer. However, we could not tell in advance which haystack the donkey would choose.

We can imagine a thought experiment in which a thousand donkeys were placed exactly between a thousand sets of haystack pairs. Although any individual donkey would remain unpredictable, we would expect that about half would turn to the right and half to the left, or that there was a 50-50 chance of a donkey picking either haystack. In statistical terminology, we would say that the chance that a donkey will eat from a given haystack is 50 percent. In each decision problem there are one or more statements or claims to be tested. Such a claim is called a *hypothesis.*

Faced with a decision problem, you may take various courses of action. For example, a typical decision problem is when a stock broker recommends that you buy a certain stock. The stock broker either implicitly or explicitly claims that he has a system for picking winners in the stock market. Suppose he claims, in fact, that his system will pick winners most of the time. He argues to your satisfaction that with this high a percentage, you will surely make money. Because the stock broker makes money in commissions, he is trying to interest you in buying a number of stocks according to his system. You are a bit skeptical of his claim and decide to test it on a few stocks which you will follow in the newspaper, but will not actually buy. You ask the stockbroker to give you the names of 10 stocks that he recommends for the purpose of seeing how many go up over the next three months. Faced with a real-life decision problem of this kind, you may take either one of two courses of action. You may wait until the end of three months and observe how many of the 10 stocks go up and then make your decision as to the broker's claim by judging these results, or you may plan in advance how many stocks must go up before you will invest your money in such a system. Both of these procedures are used in decision problems. Sometimes it is better to wait and see what happens before making a decision; at other times it is better to plan your behavior beforehand. The science of statistics covers both of these courses of action. In the first case, we use statistical procedures of data analysis; we are concerned with drawing valid conclusions from experimental data or with making proper decisions from observed facts. In the second case we use statistical theory to plan experiments so that the data is gathered with reference to the method of analysis. In any case the proper function of statistics is the making of correct conclusions or decisions from observational data. It is understood that observational data are influenced to a greater or lesser extent by uncertainty, that is, by probability factors.

Let us now develop a statistical model for making decisions. Because a decision is based on the testing of statements or claims, this part of statistics is called the testing of hypotheses. Let us now go back to our example of a stockbroker who claims he has a method of picking winners; that is, stocks that go up, say, within three months. How can we test him to see if he has the ability that he claims? We want to make our test on the basis of the 10 stocks that he picked as winners. We are skeptical and suspect that he cannot pick a winner. That is, we believe that he will just guess randomly, so that in terms of probability our belief is

$$P(\text{winner}) = P(\text{loser}) = 0.5$$

However, suppose that at the end of three months all of the 10 stocks he selected have gone up, that they are all winners. We would be much less skeptical and wonder how he does it. If we are especially stubborn, we may then ask him to give us a list of 100 stocks, and if they are all winners at the end of three months we would certainly give in and admit he has some system and would invest our money gladly, all the way. Let us analyze why we have made this decision. As we shall see this reasoning is central to decision-making.

The occurrence of 10, or even 100, correct choices by the stockbroker does not prove absolutely that his system really works. He could be guessing randomly, and the fortunate results were a streak of luck. However, on that assumption it would have to be admitted that the stockbroker is extremely lucky. It is very improbable that a person cussing at random would be right 100 times in a row. Thus if we believe that he is guessing, we must also believe that an extremely unlikely event has occurred. Note that what makes the 100 correct choices improbable is our assumption that the stockbroker is guessing. If, on the other hand, the stockbroker had inside information, and picked only stocks that were being manipulated (as some were in the 1920's), we would assume that he would always be correct, and therefore the raw data of 100 correct choices would not be improbable.

In any case, the raw data are made up of 100 correct choices. The results by themselves are neither probable nor improbable. They have a probability only in view of our prior belief about the probability of the stockbroker being correct. For example, if we believe P(winner) = 0.5 then the results are extremely improbable, whereas if we believe that P(winner) = 1.0 then the results are extremely probable. It seems sensible to adopt an assumption (or hypothesis) that makes the data reasonable than one that makes the data bizarre. We can think of our belief that P(winner) = 0.50 as an hypothesis about the stockbroker guessing. On the basis of this hypothesis, we would predict that he would be about 50 out of 100 times. The result that he was right 100 out of 100 times is extremely different from our prediction. Such a discrepancy is evidence that the initial hypothesis was incorrect, and we would say that his being right 100 out of 100 times was significant.

We call our original, or initial, hypothesis the *null hypothesis* and abbreviate it by H_o. We then collect data. If the data seems probable with respect to H_o , we say the data is *not significant* and accept H_o. On the other hand, if the data seems improbable with respect to H_o , we say that the data is *significant* and reject H_o. But how unlikely must the data be in order to be significant and so reject our null hypothesis? The answer to this question must be based on a judgment made by the investigator. A sufficiently small probability must be one small enough to convince the investigator that the null hypothesis is untenable. Common usage says that a sufficiently small probability is one given by 0.05, or 5 times out of 100, or 1 out of 20. Thus data that occur 1 time out of 20 when the null hypothesis is true will be regarded as sufficiently improbable so as to reject the null hypothesis. In other words, 1 time out of 20 (on the average) data will be judged as significant even when it is not. The result is a mistake, and is called a *Type I Error.*

The value 0.05 for the probability of rejection is called the *level of significance,* and is denoted by α. It is important to remember that the choice of 0.05 for α is arbitrary. It represents one value that is often used. Other values that are also used are α = 0.025 and α = 0.01. The particular value of a selected depends upon what one risks by accepting or rejecting the hypothesis.

Type I and Type II Errors

Ancient man distinguished between two types of numbers: peaceful ones like 2, 4, 6, 8 and the warlike ones that were in between. For example, if there were 6 animals and two people having equal claim on them, it was easy to give 3 animals to each and keep peace. If there were 5 animals, 2 could be given to each, but there may have been a fight over the one remaining. Thus the word "even" means flat and smooth, and an even number of anything can be divided into two piles of exactly the same height, that is, of even height. The even number is one that has the property of equal shares. The word "odd" comes from a word meaning "pointed." An odd number of anything can be divided only if one pile is higher or more pointed than the other. The odd number is one that has the property of "unequal shares." The expression "odds" in betting implies the wagering of unequal amounts of money.

In mathematics it is convenient to say that if two numbers are both even, they are of the *same parity,* and if two numbers are odd, they also are of the *same parity.* However an even number and an odd number, grouped together, are of *different parity.* The usefulness of this convention is based on the following:

The sum of two even numbers is even.
The sum of two odd numbers is even.
The sum of an even and odd number is odd.
The sum of an odd and even number is odd.

Using the symbols E for even and O for odd, these are
$E + E = E$
$O + O = E$
$E + O = O$
$O + E = O$

Instead of four statements, the concept of parity enables us to say the same in two statements:

Same parities add to even.
Different parities add to odd.

When, in any situation, same parities always yield one result and different parities yield the opposite result, we say that parity is conserved.

Let us now apply this convention to *decision theory*. Whenever we consider a hypothesis and a decision, the hypothesis can be either true or false and the decision can be either accept or reject. We pair a hypothesis and a decision. If the hypothesis is true and the decision is acceptance, we say they are of the same parity. Likewise if the hypothesis is false and the decision is rejection, they also are of the same parity. However, a true-hypothesis and reject-decision are of different parity. Likewise a false-hypothesis and a accept-decision are of a different parity. Thus whenever we consider an hypothesis and a decision there are four possible outcomes:

A true hypothesis can be accepted.
A false hypothesis can be rejected.
A true hypothesis can be rejected.
A false hypothesis can be accepted.

Using the symbols T for true, F for false, A for accept, and R for reject, these statements are

$T + A = $ No error
$F + R = $ No error
$T + R = $ Error (Type I)
$F + A = $ Error (Type II)

Instead of these four statements, we can say
Same parities result in no error.
Different parities result in error.

As we have stated previously a Type I Error is the case in which the hypothesis is true but we reject it. A Type II Error, then is the case in which the hypothesis is false but we accept it. Notice that we can make at most one of these two errors in any decision.

Exercises

1. A null hypothesis is rejected if
(a) it makes the probability of the data adequately high
(b) it makes the probability of the data adequately low
(c) the data is insignificant
(d) no data is available

2. Assuming the null hypothesis is true the probability of rejecting the hypothesis is
(a) 50 – 50
(b) α
(c) $1 - \alpha$
(d) 0

3. Assuming the null hypothesis is true, significant data (at the 5 percent level) is
(a) unlikely
(b) likely
(c) 95 percent likely
(d) 100 percent likely

4. If we reject the null hypothesis, we cannot make a
(a) correct decision
(b) incorrect decision
(c) Type I Error
(d.) Type II Error

5. On the basis of an experiment, it was decided that a drug claimed to be harmless produced significant side effects. A Type I Error was committed if
(a) Later definitive experiments showed the drug was extremely bad.
(b) Later definitive experiments showed the drug was harmless.

6. Data is called significant if, with respect to the claim (= null hypothesis), the data is
(a) very likely
(b) very unlikely

(c) neither likely nor unlikely

(d) made up of large numbers

7. After each of these situations write Type I Error, Type II Error, or no error.

(a) Accepted a false claim

(b) Rejected a true statement

(c) Accepted a true hypothesis

(d) Rejected a false statement

Chapter 14. Summary of probability theorems

"Are you sure/That we are awake? It seems to me/That yet we sleep, we dream" (*A Midsummer Night's Dream*)

Theorems of probability
Using the axioms of probability, it is possible to derive many theorems which play an important role in applications. The complement of a set A is denoted by A'. It is the set of elements in the universe U which are not in set A.

THEOREM 1. If A is any event in the universe U, then $P(A') = 1 - P(A)$

PROOF. Observe that A and A' are mutually exclusive by definition, and that $A \cup A' = U$ (that is, among them, A and A' contain all of the elements of the universal set U). Hence, we have

$$P(A \cup A') = P(A) + P(A')$$

and

$$P(A \cup A') = P(U) = 1$$

It follows that

$$P(A) + P(A') = 1$$

As a special case we find that $P(\emptyset) = 1 - P(U) = 1 - 1 = 0$, since the empty set $\emptyset$ is the complement of U.

In the so-called Bernoulli trial, the probability of success is denoted by p and the probability of a failure is denoted by q. According to Theorem 2.1, it follows that $p = 1 - q$.

Let us find the probability of the union of any two events regardless of whether they are mutually exclusive. This extension leads to the following theorem usually called the *general law of addition:*

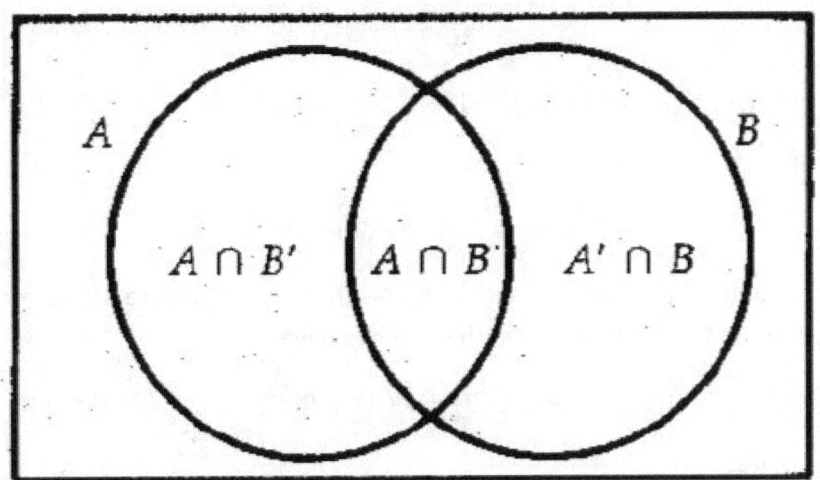

Figure 1 Partition of $A \cup B$. We see that $P(A) + P(B)$ adds the probability of $A \cap B$ twice when it should only be added once. $P(A) + P(B) - P(A \cap B)$ gives the probability of $A \cup B$.

THEOREM 2. If A and B are any events in U, then

$$P(A \cup B) = P(A) + P(B) - P(A \cap B)$$

PROOF. Observe from the Venn diagram in Figure 1 that

$$A \cup B = (A \cap B) \cup (A \cap B') \cup (A' \cap B)$$

and also that

$$A = (A \cap B) \cup (A \cap B')$$

$$B = (A \cap B) \cup (A' \cap B)$$

Since $A \cap B, A \cap B', A' \cap B$ evidently are mutually exclusive, twe have

$$P(A \cup B) = P(A \cap B) + P(A \cap B') + P(A' \cap B)$$

If we add and subtract $P(A \cap B)$, we obtain

$$P(A \cup B) =$$

$$[P(A \cap B) + P(A \cap B')] + [P(A' \cap B) + P(A \cap B)] - P(A \cap B)$$

which is

$$P(A \cup B) = P(A) + P(B) - P(A \cap B)$$

If A and B are mutually exclusive, this theorem reduces to the case when $P(A \cap B) = 0$. Thus in this case we have

$$P(A \cup B) = P(A) + P(B)$$

Exercises

1. For any set A, we have the relation $A \cup A' = U$. For any two sets A and B establish the decomposition of B given by

$$B = U \cap B = (A \cup A') \cap B - (A \cap B) \cup (A' \cap B)$$

Since $A \cap B$ and $A' \cap B$ are disjoint, show that

$$P(B) = P(A \cap B) + P(A' \cap B)$$

Recall that U denotes the universal set, and that $\emptyset$ denotes the empty set. As a first application, set $B = U$. Recall that $P(U) = 1$. Thus conclude that $P(A') = 1 - P(A)$. In particular, $P(\emptyset) = 1 - P(U)$, so that $P(\emptyset) = 0$. As a second application, suppose that $A \subset B$. Then $A \cap B = A$ and hence show that $P(B) = P(A) + P(A' \cap B)$ if $A \subset B$. Since $P(A' \cap B) \geq 0$ show that $P(A) \leq P(B)$ if $A \subset B$

Conditional probability and independence
The formal definition of conditional probability is:

DEFINITION. If A and B are any events in U and $P(B) \neq 0$, the conditional probability of A given B is

$$P(A|B) = \frac{P(A \cap B)}{P(B)}$$

If $P(B) = 0$ the conditional probability of A given B is undefined.

As a consequence of the definition of conditional probability we have the following theorem, usually called the general law of multiplication:

THEOREM 3. If A and B are any events in U, then

$$P(A \cap B) = \begin{cases} P(B)P(A|B), & \text{if } P(B) \neq 0 \\ P(A)P(B|A), & \text{if } P(A) \neq 0 \end{cases}$$

PROOF. The first relation is obtained from the definition by multiplying both sides by $P(B)$. The second relation follows from the first by interchanging the letters A and B and using the fact that $P(A \cap B) = P(B \cap A)$.

The definition of conditional probability can be motivated by the relative frequency interpretation of probabilities. Consider an experiment that is repeated a large number of times. Let the number of times the events A, B, and $A \cap B$ occur in n trials of the experiment be denoted by $N_n(A)$, $N_n(B)$, and $N_n(A \cap B)$ respectively. For n large we expect that $N_n(A)/n$, $N_n(B)/n$ and $N_n(A \cap B)/n$ should be close to $P(A)$, $P(B)$, and $P(A \cap B)$ respectively. If now we just record those experiments in which B occurs then we have $N_n(B)$ trials in which the event A occurs $N_n(A \cap B)$ times. Thus the proportion of times that A occurs among these $N_n(B)$ experiments is $N_n(A \cap B)/N_n(B)$. Since

$$\frac{N_n(A \cap B)}{N_n(B)} = \frac{N_n(A \cap B)/n}{N_n(B)/n}$$

we see that as n becomes infinite we obtain the conditional probability $P(A \cap B)/P(B)$.

Suppose a box contains N red balls labeled $1, 2, \cdots, N$ and M black balls labeled $1, 2, \cdots, M$. Assume that the probability of drawing any particular ball is $1/(M + N)$. If the ball drawn from the box is known to be red, what is the probability that it was the red ball labeled 1? Another way of stating this problem is as follows. Let B be the event that the selected ball was red, and let A be the event that the selected ball was labeled 1. The problem is then to determine the "conditional" probability that the event A occurred, given that the event B occurred. Since there are $M + N$ points each of which carries the probability $1/(M + N)$, we see that $P(B) = N/(M + N)$ and $P(A \cap B) = 1/(M + N)$. Thus

$$P(A|B) = \frac{1}{N}$$

This should be compared with the "unconditional" probability of A; namely, $P(A) = 3/(M + N)$.

PROBLEM. Compute the probability that a randomly selected item is of first grade if it is known that 4 percent of the entire production is defective, and 75 percent of the non-defective items satisfy the first-grade requirements.

SOLUTION. It is given that $P(A) = 1 - 0.04 = 0.96$, $P(B|A) = 0.75$. The required probability $p = P(A \cap B) = (0.96)(0.75) = 0.72$.

Consider a box having four distinct balls a, b, c, d and an experiment consisting of selecting a ball from the box. We assume that the balls are equally likely to be drawn, so we assign probability 1/4 to each point, a, b, c, and d in the sample space $\{a, b, c, d\}$. Let A and B be two events. For some choices of A and B, knowledge that A occurs increases the probability that B occurs. For example, if $A = \{a, b\}$ and $B = \{a\}$ then $P(A) = 1/$, $P(B) = 1/4$, and $P(A \cap B) = 1/4$. Consequently, $P(B|A) = 1/2$, which is greater than $P(B)$. On the other hand, for other choices of A and B, knowledge that A occurs decreases the probability that B will occur. For example, if $A = \{a, b, c\}$, and $B = \{a, b, d\}$, then $P(A) = 3/4$, $P(B) = 3/4$, and $P(A \cap B) = 1/2$. Hence $P(B|A) = 2/3$, which is less than $P(B)$. However, there are cases when knowledge that A occurs does not change the probability that B occurs. As an example of this, let $A = \{a, b\}$ and $B = \{a, c\}$; then $P(A) = 1/2$, $P(B) = 1/2$,

$P(A \cap B) = 1/4$, and therefore $P(B|A) = 1/2$. Events such as these, for which the conditional probability is the same as the unconditional probability, are said to be independent.

In general, if A and B are any events in a sample space, we say that A is independent of B if and only if $P(A|B) = P(A)$. Using Theorem 2.3, it can easily .be seen that if A is independent of B then B is also independent of A; that is, $P(A|B) = P(A)$ implies $P(B|A) = P(B)$ provided $P(A) \neq 0$. In such a case, it is customary to say simply that A and B are *independent*.

In the special case where A and B are independent, Theorem 2.3 leads to the following, which is usually called the *special law of multi*plication:

THEOREM 4. If A and B are independent events, then

$$P(A \cap B) = P(A)\, P(B)$$

For example, the probability of getting two heads in two successive flips of a balanced coin is $(1/2)(1/2) = 1/4$. The probability of drawing two kings in succession from a standard deck of 52 playing cards is $(4/52)(4/52) = 1/169$, provided the first card is replaced before the second is drawn. The special law of multiplication can be extended to apply to more than two independent events: if three or more events are mutually independent, the probability that they will all occur is given by the product of their respective probabilities. Thus the probability of four heads in four successive flips is $(1/2)^4 = 1/16$.

In dealing with more than two events, we must be careful; for in order for the special law of multiplication to hold, the events must be *mutually independent*. Let us give an example. We can consider a problem for three sets A, B, and C. Take the universal set a, b, c, d and assign probability $1/4$ to each point. Let $A = \{a, b\}$, $B = \{a, c\}$, and $C = \{a, d\}$. Then it can be shown that the pairs of events A and B, A and C, and B and C are independent. We say that the events A, B, and C are pairwise independent. On the other hand, $P(C) = 1/2$ and $P(C|A \cap B) = 1$. Thus knowledge that the event $A \cap B$ occurs increases the odds that C occurs. In this sense, the events A, B, and C fail to be mutually independent. In general, three events A, B, and C are mutually independent if they are pairwise independent and if $P(A \cap B \cap C) = P(A)P(B)P(C)$.

Furthermore it can be shown that if A, B, and C are mutually independent and $P(A \cap B) \neq 0$, then $P(C|A \cap B) = P(C)$.

In general we define n events $A_1, A_2, \cdots, A_n$ to be mutually independent where $n > 2$ if

$$P(A_1 \cap A_2 \cap \cdots \cap A_n) = P(A_1)P(A_2) \cdots P(A_n)$$

and if sub-collections containing at least two but fewer than n events are mutually independent.

PROBLEM. A break in an electric circuit occurs when at least one out of three elements connected in series is out of order. Compute the probability that a break in the circuit will not occur, given that the elements may be out of order with the respective probabilities 0.3, 0.4, and 0.6. How does the probability change if the first element is never out of order?

SOLUTION. The required probability equals the probability that all three elements are working. Let A_k (for $k = 1,2,3$) denote the event that the kth element functions. Then $p = P(A_1 \cap A_2 \cap A_3)$. Since the events may be assumed to be independent,

$$p = P(A_1)P(A_2)P(A_3) = (0.7)(0.6)(0.4) = 0.168$$

If the first element is not out of order, then

$$p = P(A_2)P(A_3) = (0.6)(0.4) = 0.24$$

Let us summarize. Two events A and B are said to be independent if $P(A \cap B) = P(A)P(B)$. Three events A, B, and C are said, to be mutually independent if

$$P(A \cap B \cap C) = P(A)P(B)P(C)$$

$$P(A \cap B) = P(A)P(B)$$

$$P(A \cap C) = P(A)P(C)$$

$$P(B \cap C) = P(B)P(C)$$

Thus, independence is a notion relative to a given probability measure. In contrast, the notion of disjointness is strictly a set property and does not depend upon any probability measure. In effect, events are independent if they have nothing to do with each other. For example, if we first toss a coin, then throw a die, and finally draw a card from a deck, we believe that any one of these three experiments is in no way influenced by the other two. Accordingly, in our probability model we would require that any three events A, B, C, such that A is alone determined by the coin, B by the die alone, and C by the card drawn alone, would be independent. If we apply the usual equally likely probability model, we can readily verify that the above equations for the mutual independence of A, B, C in fact do hold. For example, if A is a head, B is a 5 or 6, and C is a spade, then

$$P(A) = \frac{312}{624} = \frac{1}{2}$$

$$P(B) = \frac{208}{624} = \frac{1}{3}$$

$$P(C) = \frac{156}{624} = \frac{1}{4}$$

$$P(A \cap B) = \frac{104}{624} = \frac{1}{6}$$

$$P(A \cap C) = \frac{78}{624} = \frac{1}{8}$$

$$P(B \cap C) = \frac{52}{624} = \frac{1}{12}$$

$$P(A \cap B \cap C) = \frac{26}{624} = \frac{1}{24}$$

Exercises

1. Suppose two identical and perfectly balanced coins are tossed once.

(a) Find the conditional probability that both coins show a head given that the first shows a head.

(b) Find the conditional probability that both are heads given that at least one of them is a head. ANSWERS (a) 1/2 (b) 1/3

2. Show that if A, B, and C are three events such that $P(A \cap B \cap C) \neq 0$ and $P(C|A \cap B| = P(C|B)$, then $P(A|B \cap C) = P(A|B)$

3. If two mutually exclusive events A and B are such that $P(A) \neq 0$ and $P(B) \neq 0$, are these events independent? ANSWER. From the incompatibility of the events, it follows that $P(A|B) = 0$ and $P(B|A) = 0$; that is, the events are dependent.

4. A box has 10 balls numbered $1, 2, \cdots, 10$. A ball is picked at random and then a second ball is picked at random from the remaining nine balls. Find the probability that the numbers on the two selected balls differ by two or more.

5. The probability that the voltage of an electric circuit will exceed the rated value is p_1. For an increase in the voltage, the probability that the device will stop is p_2. Find the probability that the device will stop as a result of an increase in the voltage. ANSWER $p_1 p_2$

6. The probability that the kth unit of a computer is out of order during a time T equals p_k $(k = 1, 2, \cdots, n)$. Find the probability that during

the given interval of time at least one of n units of this computer will be out of order if all the units run independently. ANSWER. $1 - (1 - p_1)(1 - p_2) \cdots (1 - p_n)$

7. The probability that an item made on the first machine is of first grade is 0.7. The probability that an item made on the second machine is first grade is 0.8. The first machine makes two items and the second machine three items. Find the probability that all items made will be of first grade. ANSWER. 0.251.

8. A device stops as a result of damage to one tube of a total of N. To locate this tube, one successively replaces each tube with a new one. Find the probability that it will be necessary to check n tubes if the probability is p that a tube will be out of order.   ANSWER. $p(1 - p)^{n-1}$

9. The probability of the occurrence of an event in each performance of an experiment is 0.2. The experiments are carried out successively until the given event occurs. Find the probability that it will be necessary to perform a fourth experiment. ANSWER. $(1 - 0.2)^3 = 0.512$

Bayes'theorem
The general law of multiplication is useful in solving many problems in which the ultimate outcome of an experiment depends on the outcomes of various intermediate stages. Suppose we are interested in the performance of missiles received from two different suppliers, B_1 and B_2, in the proportion 3 to 1. In other words, the probability that any one missile received comes from supplier B_1 is 3/4 and the probability that it comes from supplier B_2 is 1/4. Suppose, furthermore, that 95 percent of the missiles supplied by B_{11} and 80 percent of those supplied by B_2 perform according to specifications. What we would like to know is the probability of the event A that any one missile received will perform according to specifications. We will use the relation

$$A = (A \cap B_1)(A \cap B_2)$$

We also will use the fact that $(A \cap B_1)$ and $(A \cap B_2)$ are mutually exclusive (see Figure 2).

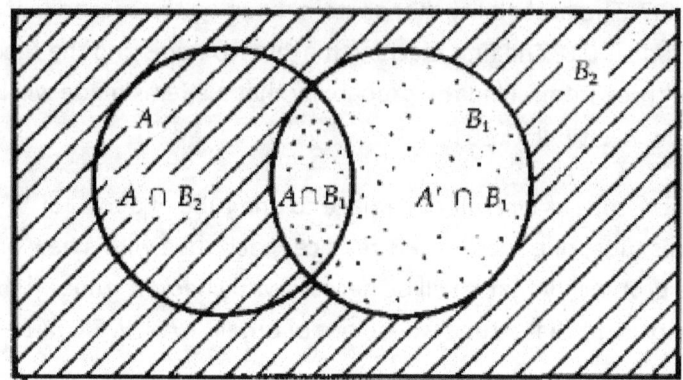

Figure 2. B_1 is the dotted area and B_2 is the shaded area.

The special law of addition yields

$$P(A) = P(A \cap B_1) + P(A \cap B_2)$$

If we apply the general law of multiplication to $P(A \cap B_1)$ and $P(A \cap B_2)$, we obtain

$$P(A) = P(B_1)P(A|B_1) + P(B_2) P(A|B_2)$$

We now substitute the given probabilities

$$P(B_1) = \frac{3}{4}, \quad P(B_2) = \frac{1}{4}, \quad P(A|B_1) = 0.95, \quad P(A|B_2) = 0.80$$

in order to obtain

$$P(A) = (3/4)(0.95) + (1/4)(0.80) = 0.9125$$

for the desired probability that any one missile received will perform according to specifications.

The above development leads to a general formula. Instead of two suppliers B_1 and B_2 at the intermediate stage, suppose there are

n mutually exclusive suppliers $B_1, B_2, \cdots, B_n$. Then the probability of the final outcome A is

$$P(A) = \sum_{i=1}^{n} P(B_i)P(A|B_i)$$

This formula for the total probability $P(A)$ is known as the *rule of elimination*. The easiest way to visualize this rule is by a tree diagram, as shown in Figure 3. The probability of the final outcome A is given by the sum of the products of the probabilities corresponding to each individual branch.

Let us now consider the so-called inverse problem. For example, suppose we have n suppliers of missiles, say $B_1, B_2, \cdots, B_n$. Suppose we want to know the probability that a particular missile came from supplier B_k when it is known that it performs according to specifications. Referring to Figure 3, we shall use the same information as before, but now we must find $P(B_k|A)$ instead of $P(A)$. To solve this problem we write the equation

$$P(B_k|A) = \frac{P(A|B_k)}{P(A)}$$

and then substitute into it the general law of multiplication

$$P(A \cap |B_k) = P(B_k)P(A|B_k)$$

and the rule of elimination

$$P(A) = \sum_{i=1}^{n} P(B_i)P(A|B_i)$$

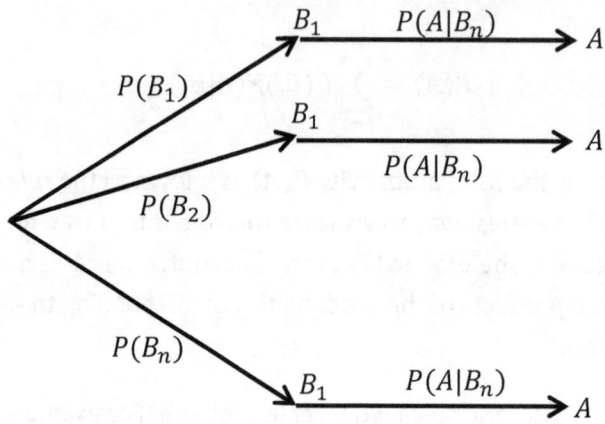

Figure. Rule of elimination

The result can be summed up in the following theorem, called *Bayes' Theorem.*

BAYES THEOREM. If $B_1, B_2, \cdots, B_n$ are mutually exclusive events of which one must occur; that is,

$$\sum_{i=1}^{n} P(B_i) = 1$$

then

$$P(B_k|A) = \frac{P(B_k)P(A|B_k)}{\sum_{i=1}^{n} P(B_i)P(A|B_i))}$$

For $k = 1,2,\cdots,n$

This formula finds the probability that the "effect" A was "caused" by the event B_{kr}. For example, in our illustration the required probability is that an acceptable missile was made by supplier B_k. The probabilities $P(B_i)$ are called the "prior" or "a priori" probabilities of the "causes" B. In practice it is often difficult to assign numerical values to the prior probabilities.

As an illustration of the use of Bayes' rule we consider the following classical problem. Suppose there are three chests each having two drawers.

(1) The first chest has a gold coin in each drawer.

(2) The second chest has a gold coin in one drawer and a silver coin in the other drawer.

(3) The third chest has a silver coin in each drawer.

A chest is chosen at random and a drawer opened. If the drawer contains a gold coin, what is the probability that the other drawer also contains a gold coin?

Often in this problem the erroneous answer 1/2 is given. However, this problem is easily and correctly solved using Bayes' rule once the description is deciphered. We can think of a probability space being constructed in which the events B_1, B_2, B_3 correspond, respectively, to the first, second, and third chest being selected. These events are disjoint and their union is the whole sample space since exactly one chest is selected. Moreover, it is assumed that the three chests are chosen equally likely so that $P(B_1) = P(B_2) = P(3) = 1/3$. Let A be the event that the coin observed was gold. Then, from the composition of the chests it is clear that

$$P(A|B_1) = 1, \quad P(A|B_2) = \frac{1}{2}, \quad P(A|B_3) = 0$$

The problem asks for the probability that the second drawer has a gold coin given that there was a gold coin in the first. This can only happen if the chest selected was the first chest, so the problem is equivalent to finding $P(B_1|A)$. We now can apply Bayes' theorem to compute, the answer, which is

$$P(B_1|A) = \frac{\left(\frac{1}{3}\right)(1)}{\left(\frac{1}{3}\right)(1)) + \left(\frac{1}{3}\right)\left(\frac{1}{2}\right) + \left(\frac{1}{3}\right)(0)} = \frac{2}{3}$$

As another illustration of Bayes' Theorem, let us consider the following example. There are two lots of items; it is known that all the items of one lot satisfy the technical standards and $1/4$ of the items of the other lot are defective. Suppose that an item from a lot selected at random turns out to be good. Find the probability that a second item of the same lot will be defective if the first item is returned to the lot after it has been checked.

Consider the hypotheses: H_1 that the lot with defective items was selected, and H_2 that the lot with nondefective items was selected. Let A denote the event that the first item is nondefective. By the assumption of the problem, we have

$$P(H_1) = P(H_2) = \frac{1}{2}, \qquad P(A|H_1) = \frac{3}{4}, \qquad P(A|H_2) = 1$$

Thus, using the formula for the total probability, we find that the probability of the event A will be to be defective. The probability of this event can also be found from the formula for the total probability. If p_1 and p_2 are the probabilities of the hypotheses H_1 and H_2 after a trial, then according to the preceding computations $p_1 = 3/7$ and $p_2 = 4/7$. Furthermore, $P(131M)^1 14, P(^{13}1H$

$$P(A) = \frac{1}{2}\left[\left(\frac{3}{4}\right) + 1\right] = \frac{7}{8}$$

After the first trial, the probability that the lot will contain defective items is

$$P(H_1|A) = \frac{P(H_1)P(A|H_1)}{P(A)} = \frac{\left(\frac{1}{2}\right)\left(\frac{3}{4}\right)}{\left(\frac{7}{8}\right)} = \frac{3}{7}$$

The probability that the lot will contain only good items is given by

$$P(H_2|A) = \frac{4}{7}$$

Let B be the event that the item selected in the second trial turns out

$$P(B|H_1) = \frac{1}{4}, \quad P(B|H_2) = 0$$

Therefore the required probability is

$$P(B) = \left(\frac{3}{7}\right)\left(\frac{1}{4}\right) = \frac{3}{28}$$

Exercises

1. A telegraphic communications system transmits the signals dot and dash. Assume that the statistical properties of the obstacles are such that an average of 2/5 of the dots and 1/3 of the dashes are changed. Suppose that the ratio between the transmitted dots and the transmitted dashes is 5:3. What is the probability that a received signal will be the same as the transmitted signal if (a) the received signal is a dot, (b) the received signal is a dash. ANSWER. (a) 3/4 (b) 1/2

2. Suppose that the population of a certain city is 40 percent male and 60 percent female. Suppose also that 50 percent of the males and 30 percent of the females smoke. Find the probability that a smoker is male. ANSWER. 20/38 or about 0.53

3. Consider 10 urns, identical in appearance, of which nine contain two black and two white balls each and one contains five white and one black ball. An urn is picked at random and a ball drawn at random from it is white. What is the probability that the ball is drawn from the urn containing five white balls? ANSWER. 5/32.

4. Consider 18 marksmen, of whom five hit a target with the probability 0.8, seven with the probability 0.7, four with the probability 0.6, and two with the probability 0.5. A randomly selected marksman fires a shot without hitting the target. To what group is it most probable that that marksman belongs? ANSWER. The second group.

5. 1n an urn, there are n balls whose colors are white or black with equal probabilities. One draws k balls from the urn, successively, with replacement. What is the probability that the urn contains only white balls if no black balls are drawn? ANSWER. n1/(1 ± 21 + • + n1).

6. The first born of a set of twins is a boy. What is the probability that the other twin is also a boy if, among twins, the probabilities of two boys or two girls are a and b, respectively, and among twins of different sexes the probabilities of being born first are equal for both sexes? ANSWER. 2a/(1 + a — b).

7. Consider that the probability of the birth of twins of the same sex is twice that of twins of different sexes; that the probabilities of twins of different sexes are equal in any succession; and that the probabilities of a boy and a girl are, respectively, 0.51 and 0.49. Find the probability of a second boy if the first born is a boy. ANSWER. 103/153.

8. Two sharpshooters fire successively at a target. Their probabilities of hitting the target on the first shots are 0.4 and 0.5 and the probabilities of hitting the target in the next shots increase by 0.05 for each of them. What is the probability that the first shot was fired by the first sharpshooter if the target is hit by the fifth shot? ANSWER. 5/11.

Chaper 15. **Bayesian decision making**

"O teach me how I should forget to think." (*Romeo and Juliet*)

Bayes rule for flipping tree diagrams

Suppose you have three sets of cousins all of whom live in the same neighborhood: the Smiths with two girls, the Jones's with a girl and a boy, and the Browns with two boys, which make up six cousins altogether. We assume that a person has an equal chance of coming in contact with any one of the three families, and once contact is made there is an equal chance of meeting either member of the family. These probability assumptions are summarized in the tree diagram.

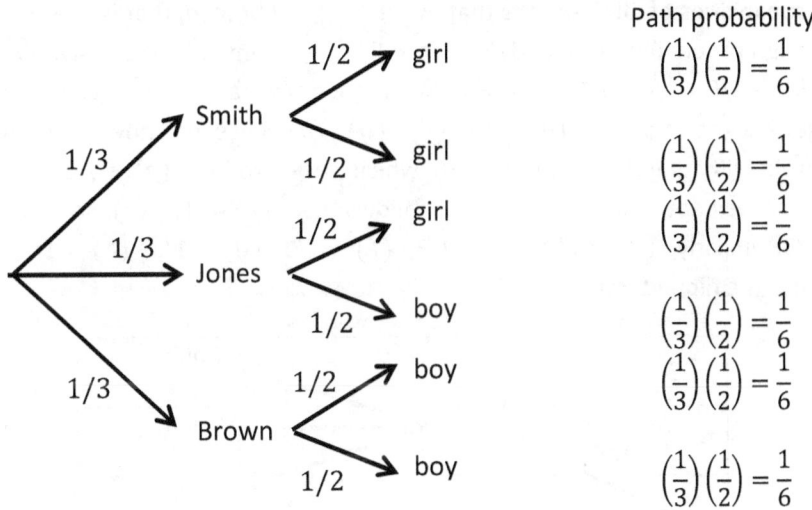

Path probability

$$\left(\frac{1}{3}\right)\left(\frac{1}{2}\right)=\frac{1}{6}$$

$$\left(\frac{1}{3}\right)\left(\frac{1}{2}\right)=\frac{1}{6}$$

$$\left(\frac{1}{3}\right)\left(\frac{1}{2}\right)=\frac{1}{6}$$

$$\left(\frac{1}{3}\right)\left(\frac{1}{2}\right)=\frac{1}{6}$$

$$\left(\frac{1}{3}\right)\left(\frac{1}{2}\right)=\frac{1}{6}$$

$$\left(\frac{1}{3}\right)\left(\frac{1}{2}\right)=\frac{1}{6}$$

On the right of this tree diagram we have computed the path probabilities, which sum to one. We now want to flip the tree diagram. A path probability in the flipped tree can be obtained by finding the corresponding path probability in the original tree. For example, the path Jones-boy in the original tree has the same probability as the path boy-Jones in the flipped tree. In this simple example, all path probabilities are the same, namely 1/6. The flipped tree diagram is

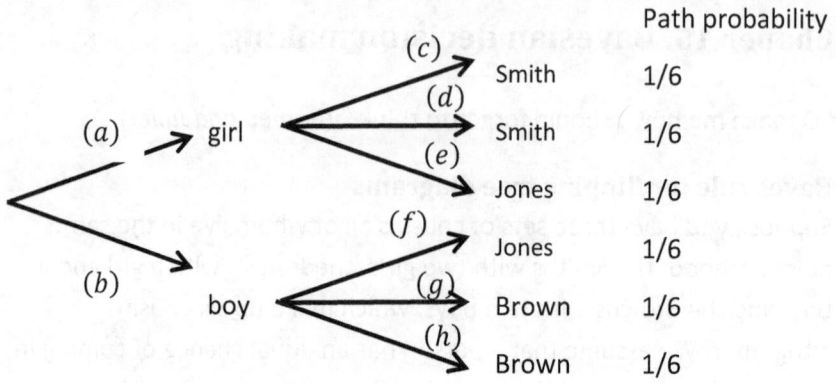

Path probability

Smith	1/6
Smith	1/6
Jones	1/6
Jones	1/6
Brown	1/6
Brown	1/6

Next we must compute the probabilities in the branches of the flipped tree. We must give the girl branch a probability equal to the sum of the probabilities of all the paths that includ   ·l branch; that is, we assign probability (a) as 1/6 + 1/6 + 1, $1/3$. Similarly, the assignment (b) is 1/6 + 1/6 + 1/6 = 1/2. Having assigned (a) and (b) we can now get assignments (c), (d), (e), (f), (g), (h). Since we know the path probability of girl and Smith is 1/6, which is the product $(a)(c)$, and since we know (a) is 1/2, it then follows that 1/6 = 1/2 (c) so (c) = 1/3. Similarly, (d) = 1/3, (e) = 1/3, (f) = 1/3, (g) = 1/3, (h) = 1/3. Thus the flipped tree is

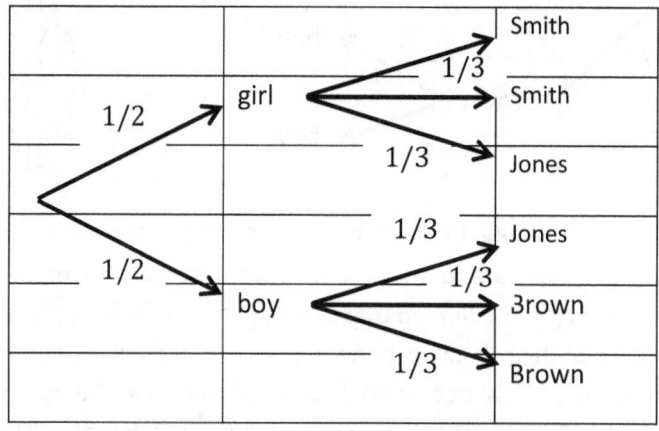

Suppose now that somebody meets one of your cousins at a party and the cousin is a girl. If he didn't catch her last name, what is the probability that it is Smith? From the flipped tree we see that from girl

there are two branches for Smith, each with probability 1/3, so the required probability is 1/3 + 1/3 = 2/3. The method we have used to find this probability is called *Bayes Rule* or *Bayes Theorem*.

Let us give another example. Suppose you are interested in all families with two children. As a matter of convenience we call a family with two girls a Smith family, a family with one of each sex a Jones family, and a family with two boys a Brown family. Suppose we meet a girl from a two-child family. From our above analysis, we have found that the probability is 2/3 that she comes from a Smith family, or in other words the probability is 2/3 that her sibling is also a girl. But we know that births are independent, and the probability of a girl is (approximately) 1/2, so the correct answer is that the probability is 1/2 that her sibling is a girl. Why has Bayes theorem given us the result 2/3 when we know the correct result is 1/2?

The answer is that Bayes theorem is correct, but our application of it to the case of two-child families is wrong. We know that the events GG, GB, BG, BB (where G = girl, B = boy) are (approximately) equally likely, so a Jones-type family which is made up of either *GB* or *BG* is twice as likely as either a Smith-type family *(GG)* or a Brown-type family *(BB)*. Thus our original tree is

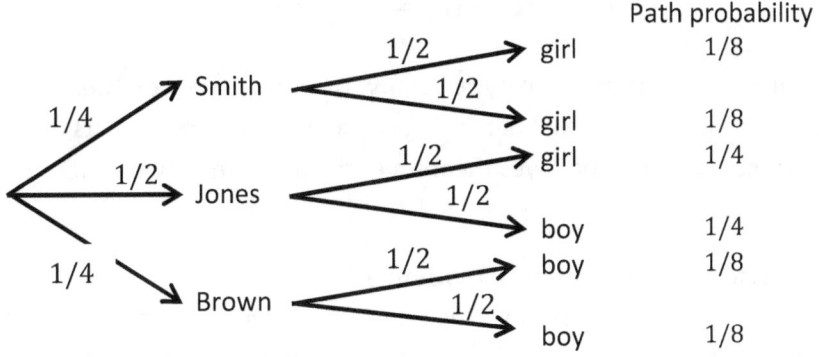

The sum of the path probabilities equals one. The flipped tree is

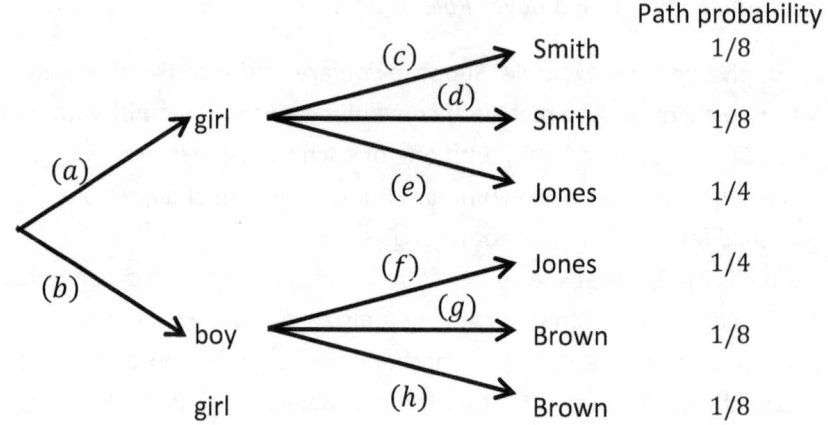

We have

$$(a) = 1/8 + 1/8 + 1/4 = 1/2$$

$$(a)(c) = \frac{1}{8} \text{ so } (c) = \frac{1/8}{1/2} = \frac{1}{4}$$

$$(a)(d) = \frac{1}{8} \text{ so } (d) = \frac{1/8}{1/2} = \frac{1}{4}$$

Hence if we meet a girl from a two-child family, the probability is

$$(c) + (d) = \frac{1}{4} + \frac{1}{4} = \frac{1}{2}$$

that she is from a Smith-type family, or in other words the probability is 1/2 that her sibling is also a girl. This represents the correct application of Bayes theorem to the problem of two-children families.

Reaction to experimental evidence
Many psychologists have investigated the intuitive reactions of subjects to experimental evidence of a probabilistic nature. Let us pose the following problem

We have two bags each with five marbles. The first bag contains three blue marbles and two white marbles, and we refer to this bag as the predominantly blue bag. The second bag contains only one blue marble and four white marbles, and we refer to this bag as the predominantly white bag. The bags are identical in appearance. One bag is drawn at random, and then one marble is drawn at random from that bag. **The marble is blue. What is the probability that it came from the predominantly blue bag?**

Various answers are given, but after a while a consensus emerges to the effect that the evidence is meager so that the probability will only be slightly better than the probability of 1/2.

We can apply Bayes rule as follows. We construct the tree diagram:

			Path probability
		3/5 → Blue marble	**3/10**
1/2 →	predominantly blue bag	2/5 white marble	2/10
1/2 →	predominantly white bag	1/5 → blue marble	**1/10**
		4/5 → white marble	4/10
			Total = 1

Given the evidence of a blue marble, what is the probability that it came from the predominantly blue bag? Instead of constructing the flipped tree diagram we can easily compute the required probability from the above tree diagram as follows. The blue marble path probabilities are shown in bold. They are 3/10 and 1/10. Their total is 4/10. The required probability is found by dividing the probability 3/10 by the total 4/10. Thus the probability of a blue bag given a blue marble is drawn is

$$\frac{3/10}{4/10} = \frac{3}{4} = 0.75$$

Let us now obtain this result by means of the flipped tree diagram:

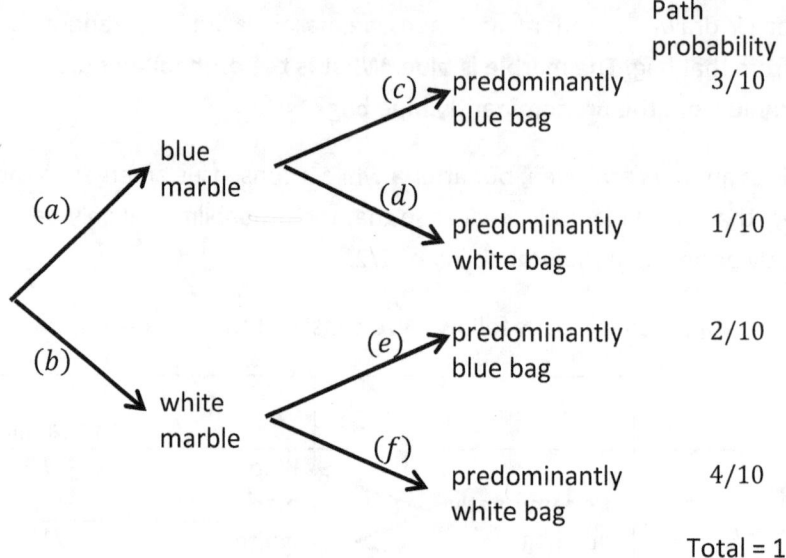

		Path probability
(c) predominantly blue bag		3/10
(d) predominantly white bag		1/10
(e) predominantly blue bag		2/10
(f) predominantly white bag		4/10
		Total = 1

We have

$$(a) = \frac{3}{10} + \frac{1}{10} = \frac{4}{10}$$

$$(a)(c) = \frac{3}{10} \quad \text{hence} \quad (c) = \frac{3/10}{4/10} = \frac{3}{4} = 0.75$$

The result 0.75 is the same as before. Thus given the evidence of a blue marble, the probability of the bag being the predominantly blue bag is 0.75. Thus the evidence has increased the probability from 0.50 (namely, the probability of drawing the predominantly blue bag) to 0.75 (namely, the probability of drawing the predominantly blue bag given the marble drawn is blue).

Hypothesis testing
A decision problem in which one of two actions must be chosen is called a problem in *hypothesis testing*. A trial in court is a problem of this type,

and the two actions are acquit or convict. Another problem in hypothesis testing is the inspection sampling of manufactured goods, where the two actions are accept or reject the lot. In such cases, one action is appropriate to certain possible states of nature, and the other action is appropriate to other possible state of nature.

The simplest kind of problem in hypothesis testing is one in which there are only two possible states of nature. For example, a child may either be well or sick; these represent the two states of nature. The two possible actions are either not call a doctor or call a doctor. A mother at home with her child must make a decision whether to call a doctor or not. Clearly the action "not call a doctor" is appropriate for the state "well" while "call a doctor" is appropriate for the state "sick." Should the mother call the doctor or should she not? To answer this question we need to know the relative cost of a wrong decision and the probabilities of the child being either well or sick. Suppose from long experience the mother knows that each time her child wakes up in the morning and says he is sick, he is actually sick only one time in four. Her prior probability of her child being sick is then 0.25 and being well is 0.75. The mother's null hypothesis is that the child is well.

If she calls the doctor when the child is really well, she is making a Type I Error. Let us assign a relative cost of 1 unit to making this Type I Error; this cost would be the doctor's fee for a service rendered that was not needed plus the psychic cost of the embarrassment of needlessly taking up the doctor's time when he is needed on other cases. This relative cost of 1 unit is called the regret r of the Type I Error.

If she doesn't call the doctor when the child is really sick, she is making a Type II Error. Let us assign a relative cost of 10 units to making this Type II Error; this cost would be the extra fees required for not catching the sickness in time plus the psychic cost of not properly treating her sick child. This relative cost of 10 units is called the regret R of the Type II Error.

These probabilities and regrets are shown in the table:

	STATE: well $P(well) = 0.75$	STATE: sick $P(sick) = 0.25$
ACTION		
Do not call doctor	0	$R = 10$
Call doctor	$r = 1$	0

We calculate the expected loss for each action by summing the losses in each row weighted by their appropriate probabilities:

Expected loss for not calling doctor = 0(0.75) + 10(0.25) = 2.50

Expected loss for calling doctor = 1(0.75) + 0(0.25) = 0.75

We see that the expected loss is less in the case of the action of calling a doctor, so the optimal decision is that action.

Up to now we have assumed that the mother has taken no statistical information as to her child's condition. The mother has available only a thermometer which she has trouble reading so there is some question as to the accuracy of her measurement. She takes her child's temperature and finds that it is 99.5° as far as she can tell. Let us suppose that temperature is normally distributed, with a mean of 99° if the child is well and with a mean of 101° if the child is sick, and with the same standard deviation equal to 1° in either case. What now is the best action, given the temperature measurement of 99.5°? At first, we might reason as follows. We draw each of the two distributions as shown in the diagram:

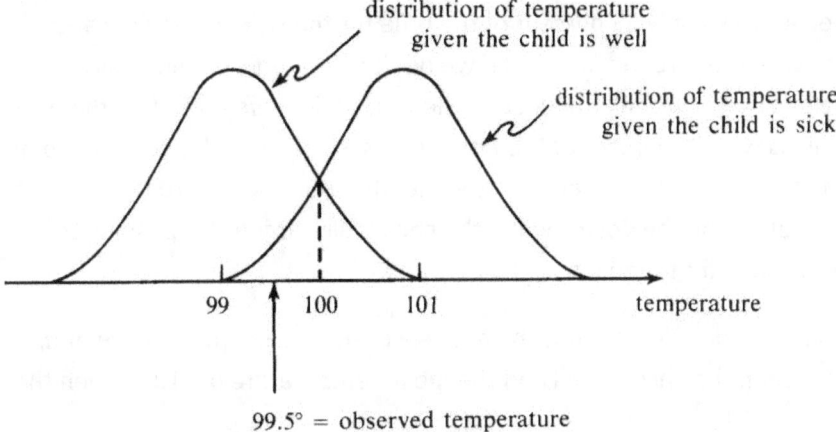

distribution of temperature
given the child is well

distribution of temperature
given the child is sick

99   100   101   temperature

99.5° = observed temperature

Using the most simple type of reasoning, we would accept the hypothesis which has the greater probability of generating the observed temperature of 99.5°. At the point 99.5° on the temperature axis, it is seen that the curve for the case of the child being well lies above the curve for the sick case, and thus on this criterion we would make the decision that the child is well and not call the doctor. In the above diagram the two curves intersect at a point corresponding to a temperature of 100°. To the left of this point the well curve lies above the sick curve, whereas to the right of this point the sick curve lies above the well curve. Hence 100° represents the decision point, and so we can formulate the following decision rule: The mother should not call the doctor if the temperature is below 100° and should call the doctor if the temperature is above 100°. The 100° mark is the break-even point, so either of the two actions would be OK at this point.

Decision Point

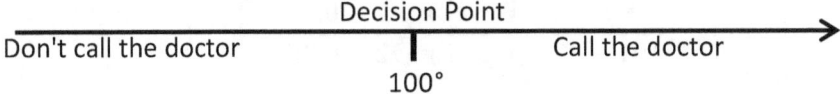

Don't call the doctor   Call the doctor

100°

All of a sudden, we say "Wait a minute!" In our analysis we have given exactly equal weighting to the two normal curves (the cases of well and sick), but we really know that 75 percent of the time the well one applies as to only 25 percent of the time for the sick one and we also know that our regret is R = 10 if we don't call when the child is sick versus r = 1 if we call the doctor when the child is well. The fact that the child is well 75 percent of the time argues for pushing the decision point to the right; on the other hand the fact that our regret is 10 times as big for not calling the doctor when the child is sick argues for pushing the decision point to the left.

The naive decision point is the average of the mean temperature of $tt_{ij} =$ 99° when the child is well and the mean temperature p = 101° when the child is sick:

$$\text{Naive decision point} = \frac{\mu_0 + \mu_1}{2} = \frac{99 + 101}{2} = 100°$$

For normal distributions with $\mu_0 < \mu$ and with a common σ, it turns out that the decision point should be moved by an amount

$$\frac{\sigma^2}{\mu_1 - \mu_0} \ln \frac{P(\text{null hypothesis true})}{P(\text{null hypothesis false})}$$

in order to take into consideration the prior probabilities, and the

$$\frac{\sigma^2}{\mu_1 - \mu_0} \ln \frac{\text{Regret of Type I Error}}{\text{Regret of Type II Error}}$$

in order to take into consideration the regrets. In our case the decision point should be moved by an amount

$$\frac{1}{101 - 99} \ln \frac{P(\text{well})}{P(\text{sick})} = \frac{1}{2} \ln \frac{0.75}{0.25} = \frac{1}{2} \ln 3$$

due to the fact that the child is well 3 times as often as he is sick, and the decision point should be moved by an amount

$$\frac{1}{101 - 99} \ln \frac{r}{R} = \frac{1}{2} \ln \frac{1}{10} = -\frac{1}{2} \ln 10$$

due to the fact that our regret for a Type II Error is 10 times our regret for a Type I Error. Here "ln" stands for the natural logarithm. A short table of natural logarithms is:

n:	1.0	1.5	2.0	2.5	3.0	4	5	6	7	8	9	10
ln n:	0.0	0.4	0.7	0.9	1.1	1.4	1.6	1.8	1.9	2.1	2.2	2.3

We must move the decision point

$$\left(\frac{1}{2}\right)\ln 3 = \left(\frac{1}{2}\right)(1.1) = 0.55$$

due to the prior probabilities and also move the decision point

$$-\left(\frac{1}{2}\right)\ln 10 = -\left(\frac{1}{2}\right)(2.3) = -1.15$$

due to the regrets. That is, we must move the decision point 0.55° to the right, and 1.15° to the left, which makes a net move of 0.55 − 1.15 = −0.60°, that is, a move of 0.60° to the left. Thus the final decision point is

$$\text{Decision Point} = \text{Naive Decision Point} + \text{Net Adjustment}$$
$$= 100° - 0.60° = 99.4°$$

The decision rule is

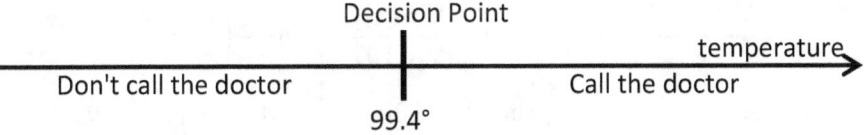

Since the mother observed a temperature of $99.5°$ she should call the doctor. In summary, suppose we have

Null hypothesis: Normal with mean μ_0 and standard deviation σ

Alternate hypothesis: Normal with mean μ_1 and standard deviation σ where $\mu_1 > \mu_0$

Prior probability of null hypothesis being true $= p$

Prior probability of null hypothesis being false $= q = 1 - p$

Regret of Type I Error $= r$
Regret of Type II Error $= R$

Then the decision point is given by

X = naive decision point
          + correction for prior probabilities and for regrets

$$= \frac{\mu_0 + \mu_1}{2} + \frac{\sigma^2}{\mu_1 - \mu_0} \ln \frac{pr}{qR}$$

and the decision rule is

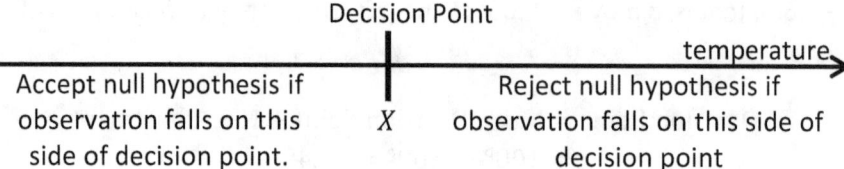

Decision Point

	X	
Accept null hypothesis if observation falls on this side of decision point.		Reject null hypothesis if observation falls on this side of decision point

temperature →

Let us compute again the decision point for our example. We have

$$X = \frac{99 + 100}{2} + \frac{1}{101 - 99} \ln \frac{0.75(1)}{0.25(10)} = 100 + \left(\frac{1}{2}\right) \ln \frac{3}{10}$$

$$= 100 + \left(\frac{1}{2}\right) \ln 3 - \left(\frac{1}{2}\right) \ln 10 = 100 + \left(\frac{1}{2}\right) 1.1 - \left(\frac{1}{2}\right) 2.3 = 99.4$$

The decision point may be interpreted as follows. We multiply the normal curve in the case the child is well by

$$P(\text{well})P(\text{Regret for Type I Error}) = pr = (0.75)(1) = 0.75$$

and we multiply the normal curve in the case the child is sick by

$$P(\text{sick})\, P(\text{Regret for Type II Error}) = (1 - p)R = (0.25)(10) = 2.5$$

Then the decision point is that value of temperature corresponding to the point of intersection of the two new curves, as seen in the diagram:

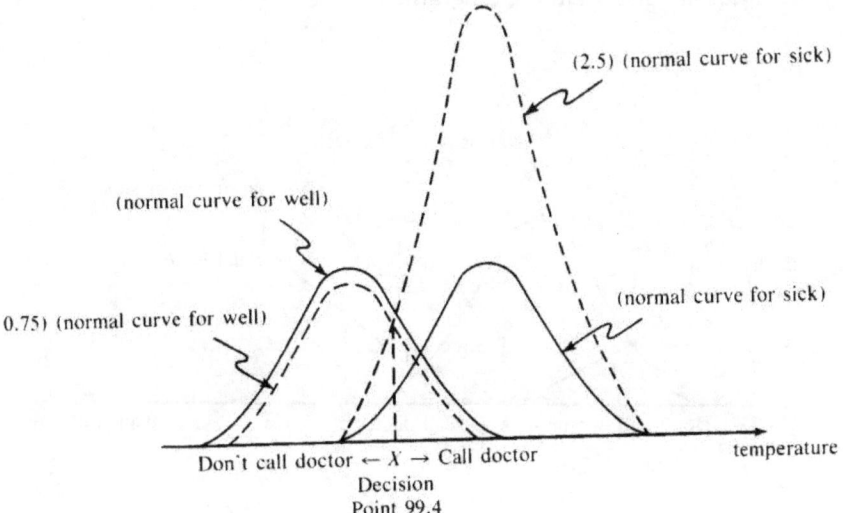

(2.5) (normal curve for sick)

(normal curve for well)

0.75) (normal curve for well)

(normal curve for sick)

Don't call doctor ← X → Call doctor

Decision
Point 99.4

temperature

The probability of a Type I Error is the area under the normal curve for well to the right of the decision point X. This probability is

$$\alpha = P\left(z > \frac{X - \mu_0}{\sigma}\right) = P(z > 0.4) = 0.34$$

The probability of a Type II Error is the area under the normal curve for sick to the left of the decision point X. This probability is

$$\beta = P\left(z < \frac{X - \mu_1}{\sigma}\right) = P(z < -1.6) = P(z > 1.6) = 0.05$$

These areas are shown in the diagram:

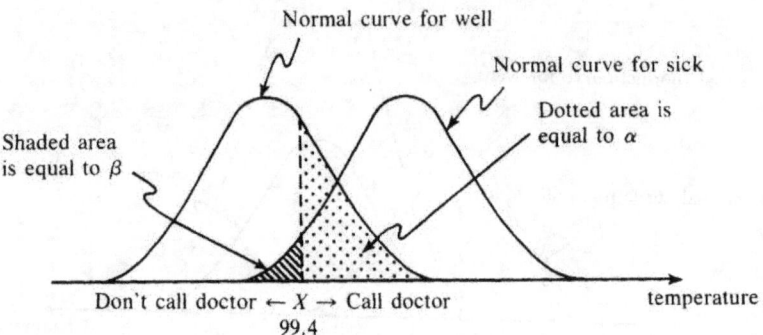

The tree diagram is

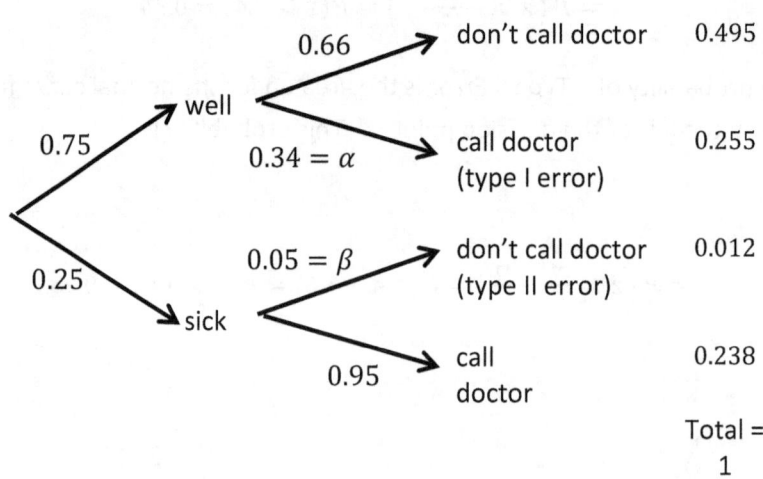

The flipped tree diagram is

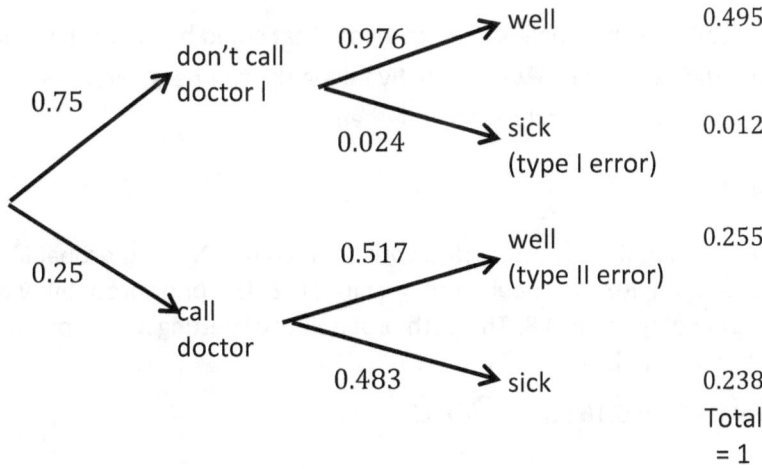

Before the mother took her child's temperature there was a 25 percent chance he was sick (i.e. the prior probability $P(\text{sick}) = 0.25$) as seen on the original tree diagram. After the mother took her child's temperature and found that the observed temperature of 99.5° fell in the critical region to the right of the decision point of 99.4°, her action would be to call the doctor. From the flipped tree diagram we see that probability that the child is sick given that this action is taken is now 0.483. In other words the posterior probability of sickness, namely 0.483, is almost twice as great as the prior probability of sickness, namely 0.250. That is the observed temperature of 99.5° was responsible for revising the probability of sickness from 0.25 up to 0.483.

Finally let us compute the expected loss. The path probability for a Type I Error is 0.255 and the regret for a Type I Error is 1. The path probability for a Type II Error is 0.012 and the regret for a Type II Error is 10. Hence the expected loss is

$$(0.255)(1) + (0.012)(10) = 0.375$$

We recall that the smallest expected loss that could be incurred without taking the temperature was 0.75. By taking the temperature we see that the expected loss is reduced by one-half.

Exercises

1. The probability of getting tails with a fair coin is 0.5. The probability of getting tails with a particular unfair coin is 0.2. The prior probability of the coin being fair is 0.8. The path probability of getting a fair coin and then getting tails is

   (a) 0.40 (b) 0.16 (c) 0.04 (d) 0.50

2.Flip the tree

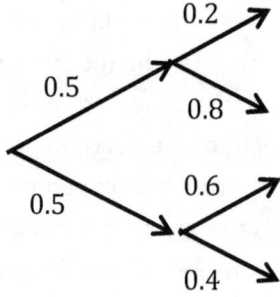

3. In a course, students were either silent S or responsive R. Seventy percent of the S students got an A while 30 percent of the R students got an A. Fill in as much as you can on the tree diagram.

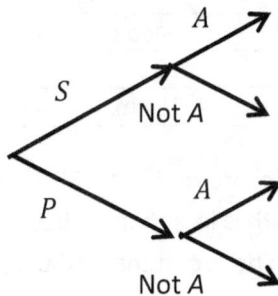

4. If, in the above problem 45% were S and 55% were R, then what is the probability that a silent student gets an A?

5. In the above problem, which is the probability that an A student is silent? (First draw the flipped tree.)

6. If the regret for a Type II Error is much greater than for a Type I Error, then we should favor calling the doctor over not calling him. TRUE or FALSE? [Here assume that the null hypothesis is well.] First identify the Type I and Type II Errors (i.e. which is "calling when well" and "not calling when sick".)

7. Suppose it rains 40 percent of the days and shines 60 percent of the days. On rainy days the barometer erroneously predicts shine 20 percent of the time, and on shiny days the barometer erroneously predicts rain 30 percent of the time. After looking at the barometer and seeing it predict rain, what is the posterior probability of rain?

(a) 0.40 (b) 0.36 (c) 0.84 (d) 0.64

8. An archaeologist has to classify skulls as Tribe A or Tribe B on the basis of their width. The populations of skull widths are normally distributed as follows:

	Mean	Standard Deviation
Tribe A	12 cm.	2 cm.
Tribe B	15 cm.	2 cm.

At first the archaeologist computed (12+15)/2=13.5, and classified a skull as Tribe A if its width is less than 13.5 and as Tribe B if its width is greater than 13.5. Then he found that Tribe A was a much more numerous tribe than was Tribe B and so he should move his decision point from 13.5 to (pick best answer):

(a) 14.5 (b) 12.5 (c) 13.5 (i.e. the same) (d) 12.0

9. A miser has great regrets if he calls the doctor when he is well but small regrets if he doesn't call the doctor when he is sick. An ordinary person might call the doctor only if his temperature is more than 99.5° but the miser would call only if his temperature was more than (pick best answer):

(a) 98° (b) 98.6° (c) 99.5° (d) 101°

Chapter 16. Decisions under uncertainty

"Wisely and slow; they stumble that run fast." (*Romeo and Juliet*)

Conditional analysis and the payoff table

The relevant consequences of any business decision, whether monetary or nonmonetary, always lie in the future; nothing we can do today will change the past. An essential part of the analysis of any decision problem, therefore, is a forecast. Difficult as it is to get accurate historical data, forecasts are even more likely to be in error. It is a rare occasion when events materialize exactly as planned. One way to deal with the problem of uncertain forecasts is to make an analysis of a decision problem on the basis of several alternative estimates of the unknown events. If, for example, we are considering the introduction of a new product which will entail an increase in fixed costs, we might calculate the profitability of the new product for several sales volumes. Sometimes such an analysis points to a single, definite decision. It might turn out, say, that even if the proposed new product had the highest sales volume considered of all possible, its contribution would still not cover the increment in fixed costs. In this case uncertainty presents no difficulties; while the "true" sales volume is not known, it is known that, no matter what this true volume is, introduction of the product would be an unprofitable decision. More commonly, it will turn out that one act would be preferable if some events occur; while some other act or acts would be better if other events occur. The decision-maker, not knowing what event will in fact occur, must somehow balance one act off against another to reach an unequivocal decision.

This chapter will be concerned with the selection of a criterion. We will examine the two major questions raised by the presence of uncertainty in a decision problem:

(1) What is a reasonable criterion for choosing among acts when their consequences cannot be predicted with certainty?

(2) Rather than making an immediate choice, should additional information be obtained in order to improve the forecasts of consequences?

A retailer wishes to decide how many units of a particular commodity to order. Since the commodity is perishable and cannot be kept in stock for more than a day, the retailer does not want to order more than a day's supply. On the other hand, since each unit costs the retailer only $1 while he sells it for $5, each demand he fails to satisfy as a result of under-ordering represents a loss to him of $4 in gross margin; to say nothing of possible loss in customer good will. If the retailer knew for certain exactly what his demand would be on a particular day, it is clear that he would want to order exactly enough units to meet the demand, no more and no less. Unfortunately, the world being what it is, the retailer does not know what demand will actually be. And yet he must decide on a particular number to stock.

In order to simplify the arithmetic in this example, we will hereafter assume that the retailer's shelf space limits him to a stock of 5 units, so that in any event he will not order more than that number. Thus his possible decisions are to stock 0, 1, 2, 3, 4, or 5 units. One possible way for the retailer to proceed is to analyze his problem under several alternative assumptions as to what his demand will be. We will call such an approach a conditional analysis because of its "if, then" character; it does not try to predict what the outcome of a particular act will in fact be, but only what it will be *conditional* on a particular event occurring. As an example, let us take the event "demand for three units" and determine what the retailer's gross profit would be from each possible act, or stock level. First of all, we note that the retailer's total cost will be $1 per unit times the number of units ordered. His revenues will be $5 x 3 = $15 providing he stocks at least three units; otherwise it will be $5 times the number of units stocked, with the remaining demands being unsatisfied. We can summarize this situation in the following table:

TABLE 1. Gross Profit if Demand is for Three Units

| | Act (Stock Level) | | | | | |
	0 item	1 item	2 items	3 items	4 items	5 items
Revenue	$0	$5	$10	$15	$15	$15
Cost	$0	$1	$2	$3	$4	$5
Gross Profit	$0	$4	$ 8	$12	$11	$10

There is no real difference between the conditional analysis of a decision problem under uncertainty and the usual economic analysis of a decision problem "under certainty." Only cost and revenues which actually depend upon what act is selected should be considered in the analysis. This means that we look only at future costs and revenues, ignoring "sunk" costs. In fact, even the future costs or revenues (which would be the same regardless of the act selected) need not be considered. In our retailer example, for instance, we can ignore any costs and revenues other than those listed in Table 1 if we can validly make the assumption that all other costs and revenues would be unaffected by the retailer's choice of a stock level. We will in fact make this assumption in our further discussions of this example, although it is unlikely to be strictly correct. The retailer's failure to meet a demand through understocking may adversely affect his sales of other items or his future sales of all items, and hence his true loss in revenue from an unsatisfied demand would be greater than $5. Again, as in the case of an economic analysis under certainty, we have included in Table 1 only those consequences which have been measured in monetary terms. While it is often possible to express noneconomic factors in terms of their monetary equivalents, this is not always practical. When important noneconomic considerations have been excluded from the formal analysis, they must be considered judgmentally in the interpretation of the analysis.

Continuing as in Table 1, we can calculate the conditional gross profit of each feasible stock level for various other levels of demand. The results of such calculations are summarized in Table 2.

TABLE 2. Payoff Table for Retailer Example

Act (Stock level)

	0 item	1 item	2 items	3 items	4 items	5 items
Demand = 0	$0	−$1	−$2	−$3	−$4	-$5
Demand = 1	0	+ $4	+ $3	+ $2	+$1	$0
Demand = 2	0	+ $4	+ $8	+$7	+$6	+$5
Demand = 3	0	+ $4	+ $8	+$12	+$11	+$10
Demand = 4	0	+ $4	+$8	+$12	+$16	+$15
Demand = 5 or more	0	+ $4	+ $8	+$12	+$16	+$20

A table such as Table 2, in which the consequences of a number of acts are evaluated for a number of events, is called a *payoff table.* Sometimes we will instead refer to the table in terms of the criterion used to evaluate the consequences; for example, we could call Table 2 a *gross profit table.* Aside from the calculation of conditional payoffs, there are two main questions involved in the preparation of a payoff table: (1) What acts should be included? (2) What events should be included?

Acts in Payoff Tables. Quite often, as in our retailer example, it will be relatively easy to list all of the feasible alternatives in a decision problem. In other problems, however, it will not be so easy to recognize what the alternatives are. No rule can be given for determining what alternatives should be considered in a particular problem. This determination must be a matter of imagination and judgment in the particular case. The best any analysis can do is to identify the best act among those examined; there might be still better acts which have not been included in the analysis.

Events in Payoff Tables. Somewhat more specific advice can be given on the events to be included in a payoff table. First of all, it should be clear from the retailer's example that the choice of events must make it possible to state in an unambiguous way the conditional payoffs of any act given each event. For example, it would not do to have "demand for 3 or 4 units" as an event in the retailer example, since it would not be possible to state unambiguously what the gross profit would be in this event if either 4 or 5 units were stocked. On the other hand, we can lump together all demands for five or more units, since the limitation on the retailer's shelf space insures that the payoff for each feasible act will be the same for all demands greater than or equal to five.

The second requirement for a listing of events is similar to the first: it must be logically impossible for two or more of the events contained in a given listing to occur simultaneously. For example, it would not do to include both "demand for 5 units" and "demand for 5 or more units" in a listing of events, since if demand is for 5 units, this occurrence cannot be assigned unambiguously to one or the other of the two events. The reason for this second requirement is to avoid the possibility of "double counting" of some of the payoffs by virtue of their being listed twice. Another way of stating that two events cannot occur simultaneously is to say that they must be *mutually exclusive.*

The third requirement for a listing of events is that it must be complete, in that some one of the events included on the list must occur. This requirement is sometimes stated by saying that the listing must be *collectively exhaustive.* This requirement is included to insure that all consequences of an act which might occur are brought into proper consideration. If, in the retailer example, we substituted "demand for 5 units" for "demand for 5 or more units," the listing would not be collectively exhaustive unless the retailer were absolutely convinced that demand would not exceed 5 units, since the payoff table would not show the consequences of demand for 6 or more units. (The fact that these consequences are the same as if demand was for 5 units does not contradict the last sentence, since it is only if demand for 6 or more units is considered specifically that we can be sure that the consequences are the same.)

If the number of acts and/or the number of events is relatively small, the payoff table is an exceptionally convenient way of organizing the economic data bearing on a decision problem under uncertainty. The student may well wonder, however, what we would do if the retailer wanted to consider any stock level up to, say, 5,000 units, and believed that his potential demand was subject to a similarly substantial range. Clearly in such a case the preparation of a payoff table would be impractical.

In a "big" problem in which the payoff table could not be constructed, one strategy would be to try to summarize the economic data in the form of algebraic equations. In the retailer's problem, for example, we could use the following equations:

$$\text{Gross Profit} = \begin{cases} 5D - S & \text{if } D \leq S \\ 4S & \text{if } D > S \end{cases}$$

where D represents the quantity demanded (event) and S represents the quantity stocked (act). These equations express exactly the same information as Table 2, since they enable us to find the gross profit for any act-event combination. However, by expressing this information algebraically, the equations are much more compact and moreover permit the determination of gross profit for many more act-event combinations than would be practical in a payoff table.

Despite the practical advantages of the algebraic form, the payoff table is still important as a conceptual tool. It is often helpful, in beginning the analysis of a decision problem under uncertainty, to imagine what the payoff table would look like, even if it is not practical to list all the acts and events. And in this note, since our objective is to bring about a conceptual understanding of decision-making under uncertainty rather than to teach techniques for "solving" particular problems, we will emphasize simple examples in which the payoff table can actually be constructed in full.

Possible criteria for selecting an act

Although the payoff table is a convenient way to organize the economic data bearing on a decision problem under uncertainty, it presents *conditional information only. We must still find a way of deciding unconditionally on a "best" act.*

One possibility is to concentrate attention on the "most likely" event. If, for example, the retailer in our example felt that demand was more likely to be for 3 units than for any other number, he might look only at the row "demand for 3" and decide to order 3 units. Concentrating on the most likely event is probably a very common way of deciding decision problems under uncertainty in practice.

Another possibility is the "conservative" approach of looking at the worst that can happen for each possible act and then choosing the act for which the "worst" outcome is the most desirable. In the retailer's problem, the most unfavorable event for any act would be "demand for 0 units." Since this event will always result in an out-of-pocket loss so long as the retailer orders any units at all, the "conservative" solution would be to order no units. Any businessman so averse to risk-taking as to favor this approach has no business being in business!

The principal drawback in each of the suggested approaches is that it concentrates on a single consequence of each act to the exclusion of all others. A fully reasoned solution of a decision problem under uncertainty ought to take into consideration all the consequences which might conceivably follow from each act. At the same time, it seems reasonable that not all of the consequences should be treated as equally important. If the retailer in our example considered the event "demand for 5 or more units" as quite unlikely, as well as considering "demand for 3 units" as most likely, then he would want to give more weight to the consequences of a demand for 3 units than to those of a demand for 5 or more units.

In the above discussion we used such phrases as "more likely" and "most likely" in referring to events. Instead of referring to likelihood in qualitative terms of "more" and "less," however, we might put likelihood on a quantitative basis by assigning numbers to events which represent their likelihood. As we know, such numbers are called probabilities.

Leaving until later the question of how probabilities might be assigned to events, let us assume that the retailer in the example we have been discussing has assigned probabilities to potential demands as given in Table 3.

The expected monetary value criterion

TABLE 3. Probability Distribution of Demand

Demand	Probability
0	0.05
1	0.15
2	0.30
3	0.25
4	0.15
5 or more	0.10
	1.00

Several paragraphs back we indicated that a good criterion for evaluating acts with uncertain consequences ought to reflect: (1) all of the possible consequences of the act; and (2) the relative likelihood of these consequences. One criterion which meets the standards just given is called expected monetary value. The expected monetary value of an act is a weighted average of the conditional monetary consequences of that act, using the probabilities assigned to events as the weights. Taking the conditional gross profits from Table 2, and the probabilities from Table 3, we can calculate the expected monetary value of the act "stock 3 units" as in Table 4.

TABLE 4. Calculation of Expected Monetary Value with Stock of 3

Demand	Probability	Conditional Profit	Weighted Profit
0	0.05	−$ 3	−$0.15
1	0.15	+$ 2	+$0.30
2	0.30	+$ 7	+$2.10
3	0.25	+ $12	+ $3.00
4	0.15	+ $12	+$1.80
5 or more	0.10	+ $12	+$1.20
Sum	1.00		+ $8.25

The total of the last column of Table 4, $8.25, is the expected monetary value of the act stock 3 units. We will also call this the *expected profit*. In other problems, we might be dealing with conditional and expected costs.

In the same manner, we can calculate the expected profits of the other stock levels open to the retailer. The results are presented in Table 5. The reader may verify one or more of these figures to insure that the method of calculation is understood.

TABLE 5. Expected Profits of all Possible Stock Levels

Stock Level	Expected Profit
0	$ 0
1	$ 3.75
2	$ 6.75
3	$ 8.25
4	$ 8.50
5	$ 8.00

The act "stock 4 units" has a higher expected profit, $8.50, than any of the other possible acts. If the retailer wishes to use the expected monetary value criterion, therefore, he would decide to stock 4 units. While the expected monetary value criterion does consider all monetary consequences of a given act and weight them according to their relative likelihoods, there are still some situations where it may not apply.

Suppose, for example, that you were offered your choice of two acts. Act 1 is absolutely certain to result in neither a profit nor a loss. Act 2, on the other hand, will lead either to a profit of $55,000 or an out-of-pocket loss of $45,000, either with a probability of one-half.

Quite possibly you would decide to pick Act 1, even though Act 2 has an expected monetary value of $5,000 versus only $0 for Act 1. There is nothing inherently "irrational" about such a choice, since there is an important difference in riskiness between the two acts which is not reflected in their expected monetary values. It may be desirable to give up $5,000 in expected value to protect against the possible loss of $45,000. Suppose, on the other hand, that you were offered a choice between Act 1 and Act 3, where Act 1 is the same as above ("do nothing") whereas Act 3 leads to either a profit of $0.55 or an out-of-pocket loss $0.45 each with probability one-half. In this case you might well be willing to act in accordance with expected monetary value and choose Act 3, since the difference in riskiness between the two acts is so slight in absolute terms.

Thus, the applicability of the expected monetary value criterion to a particular decision problem will depend upon the relative riskiness of the acts being considered. No hard and fast rule can be given, since matters will largely depend upon the circumstances of the particular decision-maker. A company with substantial financial resources, for example, will be able to assume more risk than one struggling to meet its bills; hence it may be willing to use expected monetary value in a larger number of cases. Later in this chapter we will discuss a criterion which does take into account differences in riskiness between acts: expected utility. The discussion of expected utility will also indicate a test which can always be made to ascertain whether expected monetary value is appropriate in a given decision problem.

Assessment of probabilities

Two main sets of ingredients go into the calculation of the expected monetary value of an act: (1) the conditional monetary values of the act for each possible event; and (2) the probabilities of the events.

Determination of conditional values was discussed previously; we will now be concerned with the determination of probabilities. If probabilities are to make sense when used as weights in calculating expected monetary values, they must obey certain fundamental rules. Three rules which must be obeyed by any set of probabilities are discussed in this section. In following sections we will discuss how these rules will help in assigning probabilities in particular problems.

In some event in a given decision problem is considered impossible, then the decision-maker will not want the conditional values of this event to receive any weight at all in an expected value calculation. This can be accomplished by assigning a weight (probability) of 0 to that event. If the event is considered at all possible, however, then the weight should be greater than 0. Hence we arrive at

Rule 1: A probability is a number greater than or equal to 0 assigned to an event.

This rule was used in the assignment of probabilities leading up to Table 4. It will also be observed that the probabilities in that Table add up to 1.00. There is no real logical reason why these probabilities must add up to 1.00, however; for instance, if we multiplied each probability in that Table by 3, the calculations would appear as in Table 6. Notice that, as the last step in the calculations, we must divide the weighted total of conditional profits by the sum of the weights (3.00) in order to get the weighted average of $8.25.

TABLE 6. Alternate Calculation of Expected Monetary Value with Stock of 3

Demand	Weight	Conditional Profit	Weighted Profit
0	.15	−$ 3	−$0.45
1	.45	+$ 2	+$0.90
2	.90	+$ 7	+ $6.30
3	.75	+$12	+ $9.00
4	.45	+ $ 12	+ $5.40
5 or more	.30	+$12	+ $3.60
Sum	3.00		+$24.75
Weighted Average =$24.75/3.00=$8.25			

By making the sum of the weights equal to 1.00, we can eliminate this last step in the calculations, since with weights adding up to 1.00 the weighted total and the weighted average of a set of numbers are the same. For this reason we will adopt

Rule 2. The sum of the probabilities assigned to a set of mutually exclusive and collectively exhaustive events shall be 1.

It will be noticed in Table 2 that the conditional gross profit of the act stock 3 units is the same for each of the three events demand for 3 units, demand for 4 units, and demand for 5 or more units. This being the case, it is possible to group these three events into a single event demand for 3 or more units without losing any relevant information about the act stock 3 units. This is done in Table 7.

TABLE 7. Alternative List of Events for Act 3 Units

Demand	Conditional Profit
0	−$ 3
1	+ 2
2	+ 7
3 or more	+ 12

If now we were to calculate the expected monetary value of stocking 3 units using Table 7 rather than Table 2, what probabilities should be assigned to the events? It should be evident that the events that demand is for 0, 1, or 2 units are identical with those of Table 2 and hence should get the same probabilities, i.e., those given in Table 3. Moreover, if we are to satisfy the requirements just mentioned as well as Rule 2, the weight assigned to the event demand for 3 or more units must equal the sum of the weights of the three events of which it is composed. This requirement can be generalized to

Rule 3. The probability of an event which is composed of a group of mutually exclusive events is the sum of their probabilities.

The words "mutually exclusive" in Rule 3 are necessary to avoid double-counting of some events. Consider, for example, the following probabilities derived from Table 3 by the use of Rule 3:

TABLE 8

Demand	Probability
2 or 3	0.55
3 or more	0.50

If we now want the probability of demand for 2 or more units, it would not do to add .55 and .50 to get 1.05. The difficulty is that the events demand for 2 or 3 units and demand for 3 or more units are not mutually exclusive. They contain the event demand for 3 units in common, and therefore the probability of this event is double-counted. The correct probability of demand for 2 or more units can be obtained either by: (1) applying Rule 3 to the mutually exclusive events listed in Table 3, giving a probability of .80; or (2) subtracting the probability of demand for 3 units from the total of Table 8, thereby eliminating the double-counting and also giving a probability of .80.

Probabilities based on relative frequencies

Perhaps the retailer in the example we are discussing has kept a record of the actual number of units demanded over the past 100 selling days. (Note that to do so he had to keep a record not only of actual sales but also of the number of customers turned away because of insufficient stock on hand.) The results he has obtained are summarized in Table 9.

TABLE 9. Demand History for Past 100 Days

Demand	Number of Occurrences	Relative Frequency
0	5	0.05
1	15	0.15
2	30	0.30
3	25	0.25
4	15	0.15
5 or more	10	0.10
Sum	100	1.00

We observe that the relative frequencies in Table 9, and in fact any relative frequencies, obey the following rules:

Rule 1. Each relative frequency is a number greater than or equal to 0 assigned to an event.

Rule 2. The sum of the relative frequencies assigned to a set of mutually exclusive and collectively exhaustive events is 1.

Rule 3. The relative frequency of an event which is composed of a group of mutually exclusive and collectively exhaustive events is the sum of their relative frequencies.

Except for the substitution of "relative frequency" for "probability," these rules are identical with those given for probability in the last section. As a result, the retailer might be willing to use the historical relative frequencies as his probabilities for deciding on future actions. He could argue, for example, that he considers the event demand for 3 units as being five times as likely as demand for 0 units because he has observed it five times as often in the past.

Another argument which might appeal to the retailer is based on the concept of expected monetary value. Suppose that the retailer had stocked four units on each of the past 100 days. In Table 5 we showed that this stock level led to an expected monetary value of $8.50 when used with probability weights corresponding to the relative frequencies of Table 9, and that this was the highest expected monetary value attainable. In Table 10 we calculate the profitability of such a stock level over the 100-day period.

Thus by stocking four units daily, the retailer would have made a total gross profit of $850 over the past 100 days, or an average of $8.50 per day. Both this total and this average are greater than for any other act. One justification for basing probability weights on historical relative frequencies, therefore, is that this procedure leads to choosing the act which would have been best in the period from which the frequencies were taken. This justification also points out a limitation in the use of observed relative frequencies. They are fundamentally historical data,

TABLE 10. Total Profitability of Stocking Four Units

Demand	Conditional Profit	Absolute Frequency	Profit times Frequency
0	−$4	5	−$20
1	+1	15	+15
2	+6	30	+180
3	+11	25	+275
4	+ 16	15	+240
5 or more	+ 16	10	+160
		100	+$850

and they are only relevant in making future decisions if the user is willing to make the assumption that the future will be like the past. The retailer, for example, may have just recently started to advertise the product he is selling and may believe that future demand will be greater than past demand. Even when relevant historical relative frequencies are available, the decision-maker may be unwilling to base his probability assignments solely on these frequencies. Suppose, for example, that only one selling day has elasped since the retailer began his advertising campaign and on this day five units were demanded. Should the retailer conclude on the basis of this evidence that demand for 5 units has a probability of 1.00, i.e., is certain? Common sense tells us that the answer should be "No"; observing only a single day's demand can hardly enable us to make a perfect prediction of future demands.

Judgmental weights

Even when a decision-maker lacks a firm quantitative base, in the form of relative frequencies, for assigning probabilities to events, he may have a certain amount of qualitative experience which enables him to make such judgments as "event A is more likely than event B." If these judgments could be quantified as numerical probabilities, then these probabilities could be used in calculating expected monetary values. In this section we will demonstrate how this can be done.

Reducing judgments about the likelihood of events to quantitative terms is a problem in measurement. It might help the student in understanding the procedure we are about to outline if we draw an analogy with a more familiar type of measurement. The width of this page, for example, is a quality which is susceptible to quantitative measurement. The way in which we would measure this width is to match the width of the page up against a ruler or yardstick which has previously been calibrated with certain units of measurement—say inches. The numerical width of this page is then taken as the number of units on the standard yardstick which match up against the page

In order to "measure" probability judgments, we need some sort of "yardstick" against which to match up events. We want our measurements to be such that expected monetary values can be derived from them. For such a yardstick we will use a hypothetical urn into which has been placed 100 hypothetical, serially-numbered balls (i.e., there is a ball 1, ball 2, etc., up to ball 100.) The balls have been thoroughly mixed in the urn. Now suppose that you have been offered the following opportunity: You will call out a number from 1 to 100 inclusive and then will be permitted to draw a ball from the urn without looking. If the number you call matches the number of the ball you draw, you will receive a valuable prize. Otherwise you will get nothing. Suppose that under these circumstances you would feel completely indifferent as to what number to call. For purposes of making this decision, you consider the 100 possible events resulting from drawing from the urn to be equally likely. If this judgment is to be expressed by a numerical probability, then the probabilities assigned to each event must be equal. But since, by Rule 2, the sum of these 100 probabilities must be 1.00, the probability assigned to each event must be .01. This gives us a unit of measurement for our yardstick: the probability that any one ball will be drawn from the urn. Using Rule 3, we can build up our yardstick. For example, the probability of drawing a ball with a number of 1 through 5 inclusive must be .05, since this event is composed of five mutually exclusive events each with probability .01.

Now that we have calibrated our probability yardstick, let us illustrate how it could be used to assign numerical probabilities in a real problem. We will take the event demand for 3 units in the retailer problem as an example. Suppose that the retailer has been offered a valuable prize conditional on one or the other of the two following events:

1. Demand for 3 units.
2. Drawing a ball numbered 1 through 25 inclusive from the urn.

The retailer can choose the event he prefers, but if the event he picks fails to occur he gets nothing.

Now the retailer might be of one of three frames of mind on this choice. First, he might prefer to let his prize depend upon the event demand for 3 units. That is, in his judgment demand for 3 units is more likely than drawing one of the specified balls. Alternatively, he might prefer the drawing, feeling that this event is the more likely. The third possibility is that the retailer may be indifferent between the two events, considering them equally likely. In this case they must each be assigned the same probability. But we have already seen that our probability rules require us to assign a probability of .25 to the event that one of the specified balls is drawn. Thus, in this case the retailer has succeeded in "matching up" the event demand for 3 units with a yardstick event whose probability is given, thereby arriving at a probability assignment to demand for 3 units of .25. If the two events fail to match up, so that one is preferred to the other, the yardstick event can be adjusted by changing the number of balls of which it is composed. In theory, the yardstick could be refined by using more than 100 balls. An urn with 1,000 balls, for example, could be calibrated to probabilities of .001 rather than .01.

We might wonder: What is the value of the hypothetical yardstick urn if the probabilities are to be arrived at by judgment anyway. Would it not be possible to write down the probabilities directly without going through the urn procedure? The answer to this question is "Yes," just as it is possible to write down an estimate of the width of this page without using a ruler. Proficiency at estimating widths, however, comes as a result of experience with actual measurements. Similarly, use of the yardstick urn as a hypothetical frame of reference will be helpful in developing one's intuition about probability estimates.

The value of information

One of the merits of the expected monetary value criterion is that it provides a means of determining the value of additional information. This determination is especially useful when additional information can be acquired at a cost: it will pay to acquire the information if an only if its value exceeds its cost. In this part we will show how to calculate the *expected value of perfect information* (EVPI), which is defined by:

Expected value of perfect information: The difference between the expected profit (or cost) of the optimal act under uncertainty and the expected profit (or cost) of the optimal act given complete certainty.

In other words, EVPI is the amount by which expected profit could be increased (or expected cost decreased) by the availability of a perfect forecast.

The benchmark for measuring EVPI is the profit (or cost) which would be experienced if a perfect forecast were available and action was taken accordingly. This benchmark is called *expected profit (or cost) with perfect information.* To illustrate this concept, let us return to the retailer's stock-level problem. Suppose that he were given a perfect forecast of his demand before he placed his order for the day. What would his profit be? A single, definite answer cannot be given to this question because, under uncertainty, we do not know what this forecast will be. We can, however, give conditional answers; that is, we can tell what the profit will be for any given forecast. If, for example, the forecast is that demand will be for 3 units, the retailers' best act will be to stock 3 units and this will give him a profit of $12 (see Table 2). We will call this profit the *conditional profit with perfect information* for a demand of 3 units. (In other problems, of course, we might have conditional cost with perfect information.) In a similar manner, we can get the conditional profit with perfect information for all other possible forecasts of demand. These figures are given in the column of Table 11 headed "Conditional Profit."

Even before looking at Table 11, the reader may have guessed what the next step would be: to get the expected profit with perfect information, we must multiply the conditional profits under certainty by the appropriate probabilities and add. These calculations are also indicated in Table 11. It will be noted that the probabilities assigned to the various possible forecasts are the same as those assigned to the corresponding demands in Table 3. This is because we are assuming perfect forecasts. A forecasted demand of three units, for example, would be perfect if and only if actual demand is for three units, and this demand has a probability of 0.25.

TABLE 11. Calculation of Expected Profit under Certainty

Forecast	Conditional Profit	Probability	Product
0	$ 0	0.05	$0.00
1	4	0.15	0.60
2	8	0.30	2.40
3	12	0.25	3.00
4	16	0.15	2.40
5 or more	20	0.10	2.00
		1.00	$10.40

According to Table 11, the expected profit with perfect information in the retailer problem is $10.40. The optimal act under uncertainty was shown to be stock 4 units in Table 5, and it has an expected profit of $8.50. Hence we can calculate EVPI as follows:

Expected profit with perfect information	$10.40
Expected profit under uncertainty	8.50
Expected value of perfect information	1.90

An alternative method of calculating EVPI is illustrated in Table 12.

TABLE 12. Alternative Calculation of EVPI

Demand	Probability	Conditional Profit Stock 4	Perf. Inf.	CVPI	EVPI
0	0.05	−$4	$ 0	$4	$
1	0.15	+1	4	3	0.45
2	0.30	6	8	2	0.60
3	0.25	11	12	1	0.25
4	0.15	16	16	0	0.00
5 or more	0.10	16	20	4	0.40
	1.00				$1.90

In Table 12, the column headed "CVPI" gives the Conditional Value of Perfect Information, which is simply the difference between conditional profit with perfect information and conditional profit with a stock level of 4, the optimal act under uncertainty. If, for example, the retailer gets a perfect forecast that demand will be for 3 units, he will be able to improve his profit by $1 (by stocking one less units than the 4 required under uncertainty).

In order to calculate the EVPI in a given decision problem by either of the methods just presented, it is first necessary to determine the optimal act under uncertainty. We will now illustrate another method of analysis in which the optimal act and the EVPI are determined simultaneously. To calculate EVPI, we use the profit (or cost) of action with perfect information as a reference point against which we compare the profit (or cost) of the optimal act under uncertainty. The EVPI of $1.90 calculated above, for example, indicates that the retailer's best stock level under uncertainty is expected to be $1.90 less profitable than action with perfect information. Now we could also compare other acts with action with perfect information. In Table 5, for example, it is indicated that a stock level of 3 units has an expected profit of $8.25. This is $2.15 less than the expected profit with perfect information of $10.40 given in Table 11. We can calculate this $2.15 figure in a second way, analogous to the second way of calculating EVPI given above. But first we must define the

> *Conditional opportunity loss of an act given an event:* The difference between the conditional profit (or cost) of the act and the conditional profit (or cost) with perfect information.

Conditional opportunity loss is always taken as being positive or zero. That is, it is calculated as either: (1) conditional profit with perfect information minus conditional profit of the act; or (2) conditional cost of the act minus conditional cost with perfect information. To illustrate the idea of conditional opportunity loss, the losses for the retailer example are given in Table 13; a table of this sort is called a *loss table*. The student should check a few of the values in the table by applying the above definition to the conditional profits given in Tables 2 and 11.

Probability for Business and Economics

TABLE 13. Loss Table for Retailer Example

Act (Stock Level)

Demand	0	1	2	3	4	5
0	$0	$1	$2	$3	$4	$5
1	4	0	1	2	3	4
2	8	4	0	1	2	3
3	12	8	4	0	1	2
4	16	12	8	4	0	1
5 or more	20	16	12	8	4	0

To give some meaning to the term conditional opportunity loss, let us look at the losses corresponding to a stock level of three units. If this act is chosen and actual demand turns out to be three units, the retailer will make the maximum profit he could have made given that demand; he has lost nothing. If, on the other hand, actual demand is for four units, then by stocking only three units the retailer has lost the opportunity of selling a fourth unit on which his gross profit would have been $-$1=$4; this $4 is then his conditional opportunity loss given a demand for four units. Finally, if demand is for only two units, the retailer will be overstocked by one unit. He has therefore lost the opportunity of saving the $1 cost of this unit by failing to order it.

The loss table gives conditional values only. In order to determine the optimal act, we must find the *expected opportunity loss* (EOL) of each act, i.e., the weighted average of the conditional opportunity losses using the probabilities as weights. The best act will then be the one with the lowest expected opportunity loss. The expected opportunity losses corresponding to Table 13 are given in Table 14. The student should check the calculation of at least one of these figures.

TABLE 14. Expected Opportunity Losses of All Possible Stock Levels

Stock Level	Expected Opportunity L
0	$10.40
1	6.65
2	3.65
3	2.15
4	1.90
5	2.40

Several facts may now be observed about *expected opportunity loss* (EOL):

1. The EOL of action with perfect information is, of course, always $0, since it is the reference point from which opportunity loss is measured.

2. The EOL of the optimal act under uncertainty is also the expected value of perfect information—in this case $1.90. The reason for this can be seen by comparing the definitions of conditional value of perfect information and conditional opportunity loss; for the optimal act these definitions are identical. The optimal act is the closest we can get to 0 opportunity loss on the basis of available information.

3. The EOL of all acts other than the optimal one are greater than the EOL of the optimal act. By comparing Tables 5 and 14, we can see that the difference in EOL between any two acts is equal to the difference between their expected profits. A corresponding statement would be true if we were working with cost rather than profits.

Thus, the use of expected opportunity loss as a criterion enables us simultaneously to: (1) find the optimal act; and (2) find the EVPI, which is the EOL of the optimal act.

Expected utility and attitude toward risk

Circumstances exist in which a decision-maker could feel that use of the expected monetary value criterion was inappropriate. These circumstances occur when one act has a higher expected monetary value than a second but also runs a substantially greater risk. In such a case, the decision-maker might well decide that the additional expected value of the first act is not sufficient to compensate him for its additional risk.

Previously we illustrated this point with a problem in which you were asked to choose between two acts: one leading to a "profit" of $0 with absolute certainty, and the other leading to a 50-50 chance at a profit of +$55,000 or $45,000. While the second act has an expected monetary value of $5,000, you might still prefer the first act because you wish to avoid the risk of substantial loss in the second. The amount of risk you are willing to assume is a matter of personal preference.

But while attitude toward risk is a matter of personal preference, this does not imply that it cannot be dealt with quantitatively. We have already shown how one subjective factor—relative likelihood of events— could be expressed numerically through the use of a "yardstick" urn. Somewhat more surprisingly, a very similar device can be used to reduce attitude toward risk to a quantitative basis!

Essentially, the device is to use a "yardstick" urn to assign conditional values to acts given events in such a way that the weighted averages of these conditional values are valid guides to action, even if the weighted average of conditional monetary values is not. The conditional values we will obtain are called *utilities,* and the criterion which results is called the *expected utility criterion.* Those who are familiar with the concept of "utility" as economists often use it, should be warned that it is not possible to take the weighted average of any arbitrary set of utility numbers as a guide to action. The utility numbers must be derived by a procedure which explicitly recognizes differences in riskiness.

A 50-50 chance at either $55,000 or −$45,000 might not be worth $5,000 to you, or even $0. Still, there is probably *some* amount which this chance *is* worth. Suppose, for example, that the alternative to the 50-50 chance was a certain loss of $3,750. Under these circumstances, you might very well be indifferent between the 50-50 chance and the certain loss of $3,750. We could then say that these two acts are of equivalent value, in which case the 50-50 chance is "worth" −$3,750.

Notice that this assignment of value to the 50-50 chance is entirely subjective, which is as it should be if it is to reflect personal attitude toward risk. There is no logical principle to say that the 50-50 chance is "worth" −$3,750, but only personal preference. Because of this, the assignment of a value to the 50-50 chance does not, in and of itself, help in making a better decision; it only reflects how the decision-maker would decide based on personal preferences.

Not all decisions are so easy to resolve judgmentally. It would not be so easy, for example, to assign a subjective value to the act described by Table 15. While this act has a lower expected monetary value than the 50-50 chance, intuitively it seems less risky. Our objective will be to show how choices in simple situations can be combined logically to aid the decision-maker in evaluating complex acts as this.

Instead of saying that a 50-50 chance at $55,000 or −$45,000 is "worth" −3,750, we could just as well turn this statement around and say that a certain loss of $3,750 is "worth" a 50-50 chance at $55,000 or −$45,000. While this approach may seem less natural, since we are accustomed to thinking of value in monetary terms, it will turn out to be more useful in accomplishing our purpose.

TABLE 15. Description of a More Complex Act

Consequence	Probability
$55,000	0.10
25,000	0.00
0.40	0.00
−25,000	0.15
−45,000	0.10
EMV = $ 3,500	1.00

Having shown how judgment can be used to determine that two acts have the same "value," let us now use this process to set up a "yardstick" for assigning utility values to consequences. As our "yardstick," we will use a hypothetical urn containing a number of balls, some of them marked "$55,000" and the rest marked "$45,000." You will be permitted to draw a ball unseen from the urn, and you will then receive or be required to pay the amount indicated on the ball. Before we can use this "yardstick" urn to measure utility, we will have to calibrate it, i.e., agree on a unit of measurement. Clearly, the higher the proportion of balls in the urn marked "$55,000," the more valuable the option of drawing from it. This suggests that we use the proportion of balls so marked as an index of utility value. Thus, if the urn contains only balls marked "$55,000," so that the probability of winning $55,000 is 1.00, we would say that it represented a utility value of 1.00; while if it contains no such balls, we would say that its utility value was 0.00. This assignment of values is admittedly somewhat arbitrary but will not detract from the validity of the yardstick. The situation is analogous to the measurement of temperature. In calibrating a Fahrenheit thermometer, the freezing point of water is arbitrarily assigned a value of 32° and the boiling point a value of 212°, while on a Celsius (centigrade) thermometer these points are arbitrarily assigned values of 0° and 100° respectively. Despite the arbitrariness of these assignments, both temperature scales are perfectly valid and convey the same information.

To illustrate the use of the "yardstick" urn, let us now measure the utility value of a loss of $3,750. Assuming, as we did earlier, that you would as soon incur this loss as take a 50-50 chance at $55,000 or — $45,000, these two alternatives must be assigned the same value. But, according to the way we have calibrated our "yardstick," the 50-50 chance is assigned a utility value of 0.50. Since a loss of $3,750 must be assigned the same value, its value must also be 0.50.

As a further illustration, let us assign a utility value to $0. Evidently, given that you prefer $0 for certain to the 50-50 chance, $0 is "worth" more than the 50-50 chance and so its utility value must be more than 0.50. But how much more would it be? If we represent the 50-50 chance by an urn containing 50 balls marked "$55,000" and 50 marked "–$45,000," we can increase the value of the urn by substituting "$55,000" balls for "–$45,000" balls. Eventually we may expect to be able to construct a "yardstick" urn which "matches up" against $0 for certain in the sense that you would find them equally attractive. (Here, as in the case of probability measurements, it may theoretically be necessary to increase the refinement of the measuring process by using an urn containing more than 100 balls, but this is not of great practical concern.) If this point is reached when the "yardstick" urn has, say, a 54-46 split, then we must assign a utility value of 0.54 to $0.

Proceeding in the same way, we could find the utility of any other monetary value between -$45,000 and $55,000. Of course, use of the "yardstick" urn to find the utility of every possible monetary value in every decision problem in which risk is an important factor would be so time-consuming as to be impractical. Fortunately this difficulty can be side-stepped in a fairly simple way. The decision-maker can determine the utility of a few selected monetary values, plot these values on a sheet of graph paper, and then fit a smooth curve to the plotted points. The curve can then be used to find by interpolation utilities for additional monetary values. Such a curve is illustrated in Figure 1.

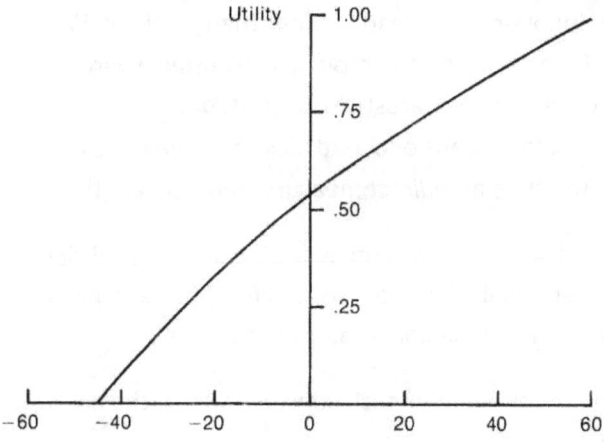

Figure 1 Utility curve, where the horizontals axis represents income (in thousands of dollars).

So far we have shown how to use a "yardstick" urn to measure the utility of any given monetary payment. However, this procedure does not in and of itself help the decision-maker to make better decisions, since the numerical measurements do no more than reflect how he would act in simple choice situations on the basis of his personal preferences. The numbers do not tell him how he should behave in these situations but only how he would behave. Once again it might help to draw an analogy with a more familiar type of measurement, as we did previously in this chapter. In measuring the width of this page with an ordinary yardstick, we could obtain a value of approximately 6 inches meaning that the width of this page matches up against 6 inch-units on the yard-stick. Similarly, the value obtained by measuring the length of this page would be approximately 9 inches. In and of themselves, these values convey only the information that the page is longer than it is wide; the number of length-units exceeds the number of width-units. By applying the rules of arithmetic to these measurements, however, we can derive new conclusions without having to use a yardstick directly. For example, we can find the width of a double-page spread: it is 12 inches obtained by addition of 6 inches and 6 inches. Thus a double-page spread is wider than it is long. We could, of course, verify this conclusion empirically by using a yardstick, but we do not need to rely on the yardstick. In short, *direct* measurements of length (by means of a yardstick) plus the rules of arithmetic enable us to arrive at *indirect* measurements of length.

In a similar way, *direct* measurements of utility (by means of a yardstick) plus the rules of arithmetic enable us to arrive at *indirect* measurements of utility. In particular, we can find the utility value of any act by:

1. Determining the utility value of each consequence of the act by direct measurement.

2. Taking the weighted average of these utility values, using the appropriate probabilities as weights.

The resulting number is called the *expected utility* of the act. The expected utility of the act described in Table 15, for example, is calculated to be 0.545 in Table 16. The utility values used in this table have been read from Figure 1. That figure in turn, you are reminded, has (by assumption) *been* constructed from direct measurements using a "yardstick" urn.

TABLE 16. Calculation of Expected Utility

Consequence	Utility Value	Probability	Weighted Utility
$55,000	1.00	0.10	0.100
25,000	0.76	.25	.190
0.00	0.54	.40	.216
−25,000	0.26	.15	.039
−45,000	0.00	.10	.000
		1.00	0.545

Since the expected utility of this act is 0.545, it is supposedly worth more than a 50-50 chance at $55,000 o   −$45,000, which has a utility value of only 0.50. The reader may well be skeptical of this conclusion, though, and so let us see whether we can verify it in another way, using direct measurement (just as we could verify indirect length measurements with a yardstick).

We can represent the act described by Table 15 by an urn containing 1,000 balls labeled as follows:

Label	No. of Balls
$55,000	100
25,000	250
0	400
−25,000	150
−45,000	100
	1,000

The balls in this urn will be thoroughly mixed and you will be allowed to draw one ball. You will then be given the amount of money corresponding to the label on the ball you have chosen. If you are willing to assume that each ball in this urn is equally likely to be drawn, then it is clear that the urn is equivalent to the act described in Table 15. Now let us bring our "yardstick" urn into the problem. To start with, we will use the "yardstick" to measure the utility of $0. In our earlier illustrations, we have assumed that you would be indifferent between

$0 for certain and a 54-46 chance at $55,000 or –$45,000. This being the case, we can put 400 balls into the "yardstick" urn of which 54%, or 216, are labeled "$55,000" and 46%, or 184, are labeled "–$45,000." Since the chance of drawing from this "yardstick" urn is worth exactly the same to you as $0 for certain, let us next *replace* the 400 balls labeled "$0" in the original urn with the 400 balls in the "yardstick" urn. We will assume that this replacement will not affect the value of drawing from the original urn, since it involves only the exchange of two things of equal value. The composition of the original urn will then be:

Label	No. of Balls
$55,000	316
25,000	250
–25,000	150
–45,000	284
	1,000

We can next perform the same operation on the balls labeled "$25,000." In Table 16, we assumed that the utility to you of $25,000, as obtained by direct measurement, was 0.76. Therefore, $25,000 is "worth" as much as the opportunity of drawing from a "yardstick" urn containing 250 balls, of which 76%, or 190, are labeled "$55,000" and 24%, or 60, are labeled "–$45,000." Substituting those 250 balls for the 250 balls labeled "$25,000" in the original urn, the composition of that urn becomes:

Label	No. of Balls
$55,000	506
–25,000	150
–45,000	344
	1,000

Finally, repeating the process with –$25,000 we can substitute for the 150 balls labeled "–$25,000" in the original urn the contents of a "yardstick" urn containing 39 (26%) balls labeled "55,000" and 111

(74%) balls labeled "–45,000." As a result, the original urn is now composed as follows:

Label	No. of Balls
$55,000	545
–45,000	455
	1,000

Since the original urn is now composed of "$55,000" and "—$45,000" balls only, it is itself a "yardstick" and we can use it to make a direct measurement of the utility of the act described in Table 15. This direct measurement, 0.545, agrees with the indirect result obtained in Table 16, thus verifying the validity of the indirect procedure.

The expected monetary value criterion discussed in this chapter is an appropriate criterion only when risk is not an important consideration in comparing acts. When risk is important, the concept of expected value can still be used, except that now it is necessary to measure the value of consequences on a "utility" scale instead of in ordinary monetary terms. The resulting criterion is called "expected utility." We have shown how a utility scale can be constructed and demonstrated its validit